Statistics for Engineers

Statistics for Engineers
an Introduction

S.J. Morrison

A John Wiley and Sons, Ltd., Publication

This edition first published 2009
© 2009, John Wiley & Sons, Ltd

Registered office
John Wiley & Sons Ltd, The Atrium, Southern Gate, Chichester, West Sussex, PO19 8SQ, United Kingdom

For details of our global editorial offices, for customer services and for information about how to apply for permission to reuse the copyright material in this book please see our website at www.wiley.com.

Library of Congress Cataloguing-in-Publication Data

Morrison, S. J.
   Introduction to engineering statistics / S.J. Morrison.
      p.   cm.
   Includes bibliographical references and index.
   ISBN 978-0-470-74556-4
   1. Engineering—Statistical methods.   I. Title.
   TA340.M67 2009
   620.0072'7—dc22

                                    2009001815

A catalogue record for this book is available from the British Library.

ISBN: 978-0-470-74556-4

Set in 10/12.5pt Times by Integra Software Services Pvt. Ltd. Pondicherry, India
Printed and bound in Great Britain by TJ International, Padstow, Cornwall

Contents

About the Author

S.J. Morrison is a Fellow of the Institution of Mechanical Engineering, the Royal Statistical Society and the Chartered Management Institute.
Morrison is also a Senior Member of the American Society for Quality.

Foreword

Jim Morrison is to be congratulated on producing this very important book. It used to be thought that to make the nearest thing to a perfect car, it was necessary for each component to be produced to satisfy the very narrowest specifications. This is the philosophy that produced the Rolls Royce. Unfortunately the car was not only exceptional in reliability but also exceptional in cost.

It is remarkable that as far back as 1957 Jim Morrison came up with a very different and important concept. This was to use in engineering design the concept of transmission of error. With this approach, it became clear that to produce low error transmission in the characteristics of an assembly, certain components had to satisfy very tight specifications and these were expensive to achieve. However, other components that had much less effect on the performance of the assembly could have much wider and less expensive specifications. He showed us how to find out which components must have very narrow specifications and which could be much less narrow. By spending money where it would do the most good, it was possible to produce a car at a moderate cost whose performance and reliability were extremely high. Morrison's concept can be applied in all areas of engineering design. His concept has had profound effects. Those companies that ignore it do so at their peril.

History is full of examples where the origin of an important concept was not known or was ignored until a much later time. This has been true in the case of robust design described above. Sometimes not only has the originator of the idea been forgotten but the essentials which he developed have been misapplied. In particular, Jim had pointed out the importance of knowing, at least approximately, the variances of the components in order to determine the variance of the assembly. In later versions of this concept such matters have been given far too little attention. We are particularly grateful, therefore, for this book in which Jim describes these techniques with clarity and accuracy.

George E.P. Box, FRS
Emeritus Professor, University of Wisconsin, USA
Honorary Member of the American Society for Quality
Inaugural holder of the George Box Medal for outstanding
contributions to business and industrial statistics awarded by the
European Network of Business and Industrial Statisticians (ENBIS)

Preface

This introductory text on statistical engineering is written by an engineer for an engineering readership. It is hoped it will appeal both to practising engineers and to students, and (indeed) to school leavers contemplating engineering as a career. It may also be useful to managers who are concerned with the quality of manufactured products. In spite of all the effort over centuries to achieve absolute precision, engineering is still (and probably always will be) beset by variability which is manifest in many different ways – properties of raw materials, the environment, measurement error, process variability, etc.

Statisticians, too, are beset by variability. If variability did not exist their branch of mathematics would (probably) never have come into existence. Variability is their focal point. They have developed powerful analytical techniques which can be of enormous benefit to society in general and to specialists in other branches of science and technology in particular.

Engineers using statistical methods need not concern themselves with profound issues of statistical inference or the subtleties of statistical mathematics. They require only familiarity with relevant statistical methods, an understanding of how they work and how to use them safely without running into danger. Some familiarity with statistical terminology is also desirable so that they can communicate with statisticians when the need arises. That is what this book is all about.

The sequence of topics is not linked in any way to the theoretical development of mathematical statistics. The text begins with a nonmathematical examination of the nature of variability in engineering data, followed by an explanation of some basic statistical methods for dealing with variability. It then follows the pursuit of variability reduction in manufacturing industry, starting with production, followed by engineering design, then research and development. Finally, measurement, statistical computing, and quality management are dealt with as background topics. Although it is convenient to use the manufacturing industry as a vehicle for demonstrating the use of statistical methods, it must be emphasised they are widely applicable in other branches of engineering.

Statistical methods provide the only satisfactory way of dealing with the variability that exists in every engineering situation. The buck doesn't stop at ground level. The responsibility for dealing with variability is carried by engineers and managers at all levels, right up to chief executive. The engineer who is lacking in statistical skill is less than competent to handle variability. For that reason statistical engineering should be a continuing professional development interest for practising engineers, irrespective of seniority.

Engineering students must recognise that statistical skills will be important to them in their future careers, no matter what branch of technology they enter or how high they set their sights. As fee-payers they are entitled to look critically at their academic curricula. If they graduate in an academic establishment at which no provision is made for teaching the elements of statistical engineering, they will find themselves later in life competing on unequal terms with statistically competent engineers who are better equipped to deal with the reality of the world.

There is a message here, too, for school leavers who are considering a professional career in engineering. They should enquire carefully about the curriculum of any engineering degree course they are thinking of entering. If there is no evidence of statistical engineering content they should pass it by and look at the next on their list before committing themselves.

This book introduces a broad range of statistical methods that are relevant to engineering. These are presented with the minimum of mathematics and the maximum of explanation. Where statistical jargon is used the words and phrases are printed in italics at the first entry so that the meaning will be self-evident from the context. The object is to build bridges of understanding between the professional disciplines of engineering and statistics.

To assist the readers who may wish to take the subject further than a basic introduction (particularly in areas of research) extensive reference lists are provided at each chapter end. In addition, four appendices offer guidance for further study. A fifth appendix accommodates statistical tables.

Lastly, by linking together the basic elements of quality – technology, management, and statistics – in a compact readable text this book should assist academic engineering schools to satisfy the requirements of the accreditation committees of professional institutions making their assessments of degree courses.

Acknowledgements

The Institution of Electrical Engineers has agreed to parts of the article 'Engineering Design – The Fount of Quality', published in the August 2000 issue of *Engineering Management Journal*, and repeated under the title 'Quality Engineering Design' in the June 2001 issue of *Manufacturing Engineer*, to be included in Chapter 4. Thanks are due to Tim Davis, Henry Ford Technical Fellow for Quality Engineering at the Ford Motor Company, for permission to use his case study in this chapter as an example of the application of variance synthesis.

Material in Chapter 7 has been reproduced from 'Quality Management' in the *Proceedings of the Institution of Mechanical Engineers* 1985, vol. 199, No. B3, 153–159 pp. by permission of the Council of the Institution of Mechanical Engineers.

The author is grateful to Elwyn Davies and Hefin Rowlands of the former IEE Quality Management Professional Group and to Clare Morris and Dan Grove of the RSS Quality Improvement Committee for their support and encouragement in organising joint discussions in 1999/2000 on statistical engineering issues between the Institution of Electrical Engineers and the Royal Statistical Society in London.

Thanks are due to Julian Booker, Simon Edwards, Allan Reese and Dave Stewardson who have all been helpful in a variety of ways. Thanks are also due to the staff of the Industrial Statistics Research Unit at the University of Newcastle upon Tyne for assistance in preparing the text for distribution on disc in CD ROM format in 2004, and also to Dan Grove for updating references to computer software in Chapter 6.

1

Nature of Variability

There is no engineering product so simple that only one source of variability affects its dimensions or properties. Take two examples of products which are relatively simple in their physical appearance – high-carbon steel wire and milk bottles.

The tensile strength of steel wire depends on numerous factors: the carbon content of the ingot from which rods were made in the rolling mill; the temperature of the heat treatment furnace through which the rods were passed; the rate of passage through the furnace; the temperature of the quenching bath; the ambient temperature in the heat treatment shop; the number of dies through which the rods were drawn to finished wire size; the rate of drawing; the ambient temperature in the wire mill, etc. Variability in any of these factors is likely to generate variability in tensile strength.

One of the hazards of a milkman's life is the possibility of being stopped in the street by a weights and measures inspector. Milk bottles are filled to a predetermined level on automatic machines. The capacity at that level is determined by the external profile of the bottle and by its wall thickness. The bottles are made on multi-head automatic machines by dropping gobs of molten glass into metal moulds (one at each work station), piercing them hollow, then inflating them with compressed air until they fill the moulds. The external profile can be affected by different settings at each work station, by mould differences, by fluctuations in air pressure, by sagging after release from the moulds, and by malfunctioning of the automatic timing gear which

Statistics for Engineers: an Introduction   S.J. Morrison
© 2009 John Wiley & Sons, Ltd

controls the various functions. The wall thickness is determined by the setting of the gob feeder and this in its turn is affected by the viscosity of the glass, the forehearth temperature, and the action of the shears which cut off successive gobs from the continuous flow of the feeder. Variability in any of these process factors may contribute to variability in the volumetric capacity of bottles in continuous production.

It must be assumed that most engineering products which are infinitely more complex than steel wire or milk bottles will be equally susceptible to a multitude of factors located in raw materials, components, processes and the environment which are capable of affecting the properties and dimensions of a finished product. It is therefore important for engineers to have an understanding of the way in which random combinations of independent sources can affect the variability of a finished product.

This can be demonstrated with random combinations of the variables R, Y and B in Table 1.1. These single-digit numbers in the range 0–9 were generated by throwing twenty-faced icosahedron coloured dice (red, yellow and blue) with the numbers zero to nine engraved twice on each die. The dice were invented in 1950/60 by Mr Yasushi Ishida and patented by Tokyo-Shibaura Electric Company. They were marketed and distributed by the Japanese Standards Association for demonstrating the principles of statistical quality control. In the discussion that follows the data in Table 1.1 will be used to demonstrate some of the phenomena of variability that are encountered in engineering data without resort to the mathematics of probability theory. It is hoped this will help the reader to understand the relevance of statistical methods to be described later.

Table 1.1 Dice scores

R	Y	B	$R+Y+B$	Mean	Range	$R \times Y$
0	6	5	11			0
0	8	9	17			0
4	6	5	15	13.8	6	24
7	0	6	13			0
9	4	0	13			36
1	9	4	14			9
7	0	3	10			0
7	3	6	16	12.2	9	21
2	4	1	7			8
1	9	4	14			9

Continued for one hundred trials

One hundred trials were conducted, but only the first ten are recorded in the table.

Readers who are not convinced that the trials are properly reported are at liberty to conduct their own time-consuming experiments. Also recorded in the table are the sums $R + Y + B$, and the products $R \times Y$, along with the *mean* and the *range* of groups of five. In statistical terms, the mean of a set of data is the sum of the individuals divided by the number of individuals. The range is the difference between the largest and smallest individuals.

The *frequency distributions* are as follows;

R, Y and B	Frequency	R + Y + B	Frequency
0	30	0, 1	0
1	38	2, 3	1
2	20	4, 5	2
3	38	6, 7	7
4	29	8, 9	12
5	31	10, 11	15
6	29	12, 13	24
7	32	14, 15	17
8	21	16, 17	4
9	32	18, 19	9
		20, 21	5
		22, 23	3
		24, 25	1
		26, 27	0

These can be represented graphically in Figures 1.1 and 1.2.

In a perfect world one might expect Figure 1.1 to display 30 scores in each of the 10 categories 0–9, but the bar chart (or *histogram*, to use a statistical term) shows some degree of irregularity. If bias was suspected it would be necessary to run a much more extensive series of trials to show whether the dice were loaded in favour of scores 1 and 3 at the expense of scores 2 and 8. In the absence of such evidence it can be assumed that the scoring conforms to a rectangular distribution and that the irregularity is no more than is commonly encountered in real life collections of data.

In sharp contrast, the bar chart for the sum of the three colours (Figure 1.2) shows an entirely different pattern of distribution. There is a marked central tendency around a mean score of 13.5 which is easy to explain. All possible combinations of scores on the three dice are equally likely. There are many

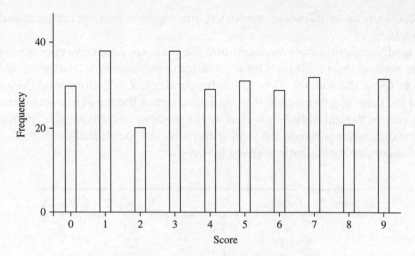

Figure 1.1   Individual dice scores

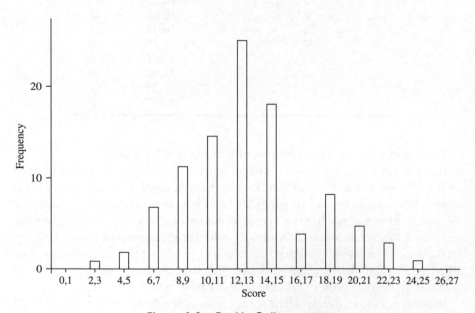

Figure 1.2 $R + Y + B$ dice scores

different combinations, yielding totals of 10, 11, 12, 13, 14 or 15, but very few which can yield extreme values of 0, 1, 2, 3 or 24, 25, 26, 27. In fact there is only one combination $0 + 0 + 0$ which could yield 0 and only one other combination $9 + 9 + 9$ which could yield 27, and neither occurred in this relatively small set of trials.

Symmetrical bell-shaped distributions exhibiting a central tendency are commonplace in engineering data. It is not unreasonable to argue these are indicative of random combination of independent factors contributing to the variability of the data and to suggest that analytical statistical methods might be used to identify and control them.

However, it must not be assumed that other patterns of distribution will not occur in engineering data. The distribution of products of red and yellow scores, $R \times Y$ is highly *skewed* (i.e. asymmetric) as shown in Figure 1.3.

Skewed distributions do occur in engineering when the effect of a contributory factor is one-sided. For example, in a thermionic valve electrons are emitted from the heated cathode and are attracted by a positive voltage on the anode. They have to pass through the grid (a helix of fine wire) to which a

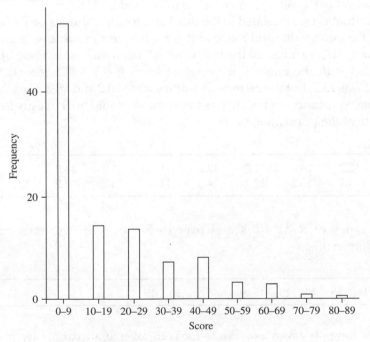

Figure 1.3 $R \times Y$ dice scores

negative voltage is applied to control the current. Any lack of uniformity in the grid helix can only increase, not reduce, the anode current. Again, in a cylindrical mechanical product zero ovality is the ultimate degree of perfection. Any finite degree of ovality is positive if it is regarded as the excess of the major diameter over the minor without regard to orientation. In such circumstances skewed data distributions are inevitable.

Fortunately statistical methods are available which are not confined exclusively to data that conforms to a symmetrical distribution. When skewed distributions are encountered in engineering data they can often be handled more easily by making a logarithmic transformation of the data.

The data in Table 1.1 can be used to demonstrate relationships between *samples* and *populations*. This is a matter of considerable importance to engineers who often have to draw valid conclusions from quite small samples of data. For example, in the early stages of development of a new product it is necessary to check measurements of a few prototypes to determine whether the population will be on target and whether the (unavoidable) spread of variability will lie comfortably within specification tolerance limits. In this instance the prototype data can be treated as a sample from a population that does not yet exist, yet a prediction has to be made.

This situation is simulated in the third and fourth columns of Table 1.1 by taking the mean value and range of $R + Y + B$ scores in successive groups of five trials. This resulted in the following 20 mean values, not one of which coincided with the mean of the original set of $R + Y + B$ scores (12.9). The nearest was 13.2, but the extreme examples were 10.2 and 15.2. Clearly, there were many instances in which the sample mean would not have given a good estimate of the population mean.

13.8	12.2	13.4	11.2	12.2	11.2	12.0	14.0	10.8	15.0
13.8	14.4	15.2	12.2	14.0	12.2	13.2	14.2	13.2	10.2

The range of $R + Y + B$ scores over each group of five trials gave the following results.

6	9	13	16	10	15	4	9	8	12	15	19	12	5	13	7	9	17	13	7

If the range is taken as a crude measure of overall variability (as many development engineers have been known to do in the past when writing specification tolerances) it is clear that not even the highest value (19) recorded in this set of trials would embrace the span of the distribution shown in Figure 1.2. Most of the others would fall very far short of this requirement.

The relatively small sets of data used by engineers at the development stage of a new product can be regarded as samples from a population which will exist when full-scale production starts. The discrepancies in mean value and variability which can exist between a sample, and the population from which it is drawn, identify a serious hazard along the road from design, through development to production of manufactured products. It is to be hoped that the straightforward demonstration of the risks given above will alert engineers to the dangers and persuade them to listen more carefully to the advice of statisticians, or (better still) develop some statistical skill on their own account. So, if range is not to be regarded as a satisfactory measure of overall variability what else can we do? Consider the following small set of data:

16	18	16	10	14

The location of the data on a scale of measurement can be identified by calculating the mean value.

$$(16 + 18 + 16 + 10 + 14)/5 = 74/5 = 14.8$$

The *deviates* of the individuals from the mean are

$$16.0 - 14.8 = 1.2$$
$$18.0 - 14.8 = 3.2$$
$$16.0 - 14.8 = 1.2$$
$$10.0 - 14.8 = -4.8$$
$$14.0 - 14.8 = -0.8$$

The sum of these deviates, taking account of positive and negative signs, will be zero. Suppose we square them before adding them together?

$$1.2^2 + 3.2^2 + 1.2^2 + (-4.8)^2 + (-0.8)^2 = 1.44 + 10.24 + 1.44 + 23.04 + 0.64$$
$$= 36.80$$

This *sum of squares* is a powerful overall measure of variability which gives equal weight to all of the individuals, not just the extreme values. It does, however, respond to the size of the data. If data from the same source had 10 values the sum of squares would be (roughly) twice as large.

This can be overcome by dividing the sum of squares by the number of individuals to give a *mean square*:

$$\frac{36.80}{5} = 7.36$$

In some situations the divisor should be one less than the number of individuals, but more of that later in Section 2.2!

Summing squares to measure variability is the foundation on which statistical analysis is built. In modern usage 'statistics' implies much more than simply recording events. In the hands of a competent engineer statistical analysis is a powerful tool which should not be neglected. Now read on!

2

Basic Statistical Methods

The absence of a sound statistical element in an engineering degree is a serious weakness. A course in quality assurance embracing techniques of applied statistics along with principles of operations management would be appropriate (Morrison, 1997). The necessary basic statistical methods are presented in this chapter. The elements of operations management are presented in Chapter 7.

2.1 Variance

Engineers wishing to make extensive use of statistical methods must first come to terms with *statistical variance*. The concept of *moments* which is familiar to engineers also serves its purpose in statistical analysis. Moments of the first order are used by engineers to determine the centre of gravity of an assembly of components. Statisticians use moments of the first order to determine the location on the scale of measurement of the mean value of a set of data. Moments of the second order are used by engineers to deal with the inertia of rotating masses. Moments of the second order are used by statisticians to measure the dispersal of individuals about the mean in a data set. This is termed *statistical variance*.

Statistics for Engineers: an Introduction   S.J. Morrison
© 2009 John Wiley & Sons, Ltd

If n individual values x_i in a set of data are represented by the symbols x_1, $x_2, x_3 \ldots x_i \ldots x_n$ then the mean $\bar{x}$ and the variance $V(x)$ are given by

$$\bar{x} = \frac{1}{n} \sum x_i$$

$$V(x) = \frac{1}{n} \sum (x_i - \bar{x})^2$$

Consider a data set of five values $1, 8, 8, 9, 6$. The mean and variance can be calculated as follows:

$$\sum x_i = 1 + 8 + 8 + 9 + 6 = 32$$

$$\therefore \bar{x} = \frac{32}{5} = 6.4$$

$$\sum (x_i - \bar{x})^2 = (1 - 6.4)^2 + (8 - 6.4)^2 + (8 - 6.4)^2$$
$$+ (9 - 6.4)^2 + (6 - 6.4)^2 = 41.20$$

$$\therefore V(x) = \frac{41.20}{5} = 8.24$$

When calculating the sum of squares of the *deviates* of the individuals about the mean it is often more convenient to use the algebraic identity

$$\sum (x_i - \bar{x})^2 = \sum x_i^2 - \left(\sum x_i\right)^2 / n$$

The quantity $\sum x_i^2$ is sometimes referred to as the *crude sum of squares*. $(\sum x_i)^2/n$ is then the *correction factor* and $\sum(x_i - \bar{x})^2$ is the *adjusted sum of squares*. Applying this procedure to the set of data above gives the same result as before:

$$\sum x_i^2 = 1^2 + 8^2 + 8^2 + 9^2 + 6^2 = 246$$

$$\left(\sum x_i\right)^2 / n = 32^2/5 = 204.80$$

$$\sum (x_i - \bar{x})^2 = 246.00 - 204.80 = 41.20$$

It is not very convenient to have the mean and the variability expressed in different units of measurement, such as 'miles per hour' and 'miles per hour squared'. To overcome this difficulty the square root of variance is termed the *standard deviation, sigma*.

$$\sigma = \sqrt{V(x)}$$

For the set of data considered above:

$$\sigma = \sqrt{V(x)} = \sqrt{8.24} = 2.87$$

The set of five values 1, 8, 8, 9, 6 can now be summarised in statistical terms as having a mean value $\bar{x} = 6.4$ and a standard deviation $\sigma = 2.87$. Note that the standard deviation (or the variance) is a powerful measure of variability taking account of every individual in the data set, not just the extreme values 1 and 9. In this respect it is superior to the range (i.e. the difference between the largest and smallest values) which is often used by engineers as a crude measure of variability. Moreover, no prior assumption is made about the shape of the parent distribution. The entire data set is taken just as it stands.

Engineers will appreciate that the statistical mean is analogous to a centre of gravity and the statistical standard deviation is analogous to the radius of gyration of a rotating mass.

It should be pointed out that statisticians sometimes use two other central values besides the mean when discussing a set of data. The *median* is the mid-point of the data when the individuals are arranged in order of magnitude. The *mode* is the most commonly occurring value. The mean, the median and the mode sometimes coincide exactly, but this is not an invariable rule.

2.2 Divisor '*n*' or '*n*−1'?

Engineers using a hand-held calculator for statistical calculations may be perplexed to find two keys labelled 'σxn' and 'σxn−1' (or corresponding subroutines in computer software). Which one should be used?

To clarify this it is necessary to consider the relationship between a sample and the larger population from which it was drawn. To make a distinction between population and sample the symbols used for mean and standard deviation will be $\bar{X}$ and σ for the population, $\bar{x}$ and s for the sample. It will be assumed that the purpose of calculating $\bar{x}$ and s from the sample data will be to estimate the unknown parameters $\bar{X}$ and σ of the population.

With the knowledge that the 1, 8, 8, 9, 6 data set in the previous section originated as five throws of an unbiased twenty-faced die, capable of generating a rectangular distribution of numbers in the range zero to nine, the data can be treated as a random sample from an infinite population with a mean value $\overline{X} = 4.5$. The true sum of squares about the population mean is therefore:

$$\sum(x_i - \overline{X})^2 = (1 - 4.5)^2 + (8 - 4.5)^2 + (8 - 4.5)^2 + (9 - 4.5)^2$$
$$+ (6 - 4.5)^2 = 59.25$$

Obviously this is greater than the sum of squares previously calculated about the sample mean:

$$\sum(x_i - \overline{x})^2 = 41.20$$

As has already been seen in the previous chapter it is quite unusual for a sample mean to coincide exactly with the population mean. In the general case, when the population mean is not known in advance, the sum of squares about the sample mean will underestimate the true sum of squares and the sample standard deviation s will underestimate the population standard deviation σ. This can be compensated by using the divisor '$n-1$' when calculating the variance. This is not just a fudge – there is sound mathematical reasoning to show this gives the best estimate of the population standard deviation σ.

Hence the rule: If the purpose is simply to calculate the variance of a set of data, use the divisor 'n', but if the purpose is to estimate the standard deviation of the population from which the sample may have been drawn, use the divisor '$n - 1$'. Obviously the difference is neither here nor there in large data sets, but there is a considerable difference with small samples which quite often crop up in engineering (for example, at the prototype development stage of a new product).

So, when estimating the variance of a population from sample data we use the expression

$$s^2 = \frac{1}{n - 1} \sum(x_i - \overline{x})^2$$

Computed in this way s^2 is an *unbiased estimator* of the population variance $V(x)$. The quantity $n-1$ is referred to as the number of *degrees of freedom* associated with the estimate. The sum of the n deviations $(x_i - \overline{x})$ is zero by virtue of the definition of the mean. If values are assigned to $n - 1$ individuals the remaining one is already determined. In many forms of statistical analysis the degrees of freedom are identified by the symbol v.

2.3 Covariance and Correlation

Engineers will sometimes encounter *bivariate* data in which two variables such as x and y appear to be correlated. *Statistical covariance* (cov) measures the degree of association, using sums of products in place of sums of squares:

$$\text{Cov}(x,y) = \frac{1}{n}\sum(x_i - \bar{x})(y_i - \bar{y})$$

where y_i is the individual value of y associated with an individual x_i.

As in the case of sums of squares, there is a useful algebraic identity for simplifying the calculation of *sums of products*:

$$\sum(x_i - \bar{x})(y_i - \bar{y}) = \sum x_i y_i - \frac{1}{n}\sum x_i \sum y_i$$

A graphical interpretation of covariance is given in Figure 2.1, where individuals are plotted on an X, Y coordinate field.

Since we are considering the product of the x and y deviates from their means it is appropriate to use an origin at the centroid of the data with axes representing the deviates $(x_i - \bar{x})$ and $(y_i - \bar{y})$. These divide the field into four quadrants, upper right, upper left, lower left, lower right. The products will be positive in the upper right and lower left quadrants. They will be negative in the upper left and lower right quadrants.

If there is no association between the x and y variates, as in Figure 2.1(a), the positive and negative products will cancel out. If there is a strong association then either the positive products will predominate, as in Figure 2.1(b), or the negative products, as in Figure 2.1(c).

A dimensionless *correlation coefficient r* can be used to measure the degree of association:

$$r = \frac{\sum(x_i - \bar{x})(y_i - \bar{y})}{\sqrt{\sum(x_i - \bar{x})^2 \sum(y_i - \bar{y})^2}}$$

If there is a perfect association between the x and y variates as in Figure 2.1(d) the square of the sum of products will be numerically equal to the product of the sums of squares and the correlation coefficient will be unity. It will be positive for a rising gradient and negative for a falling gradient, depending on whether the products of the x and y deviates are positive or negative.

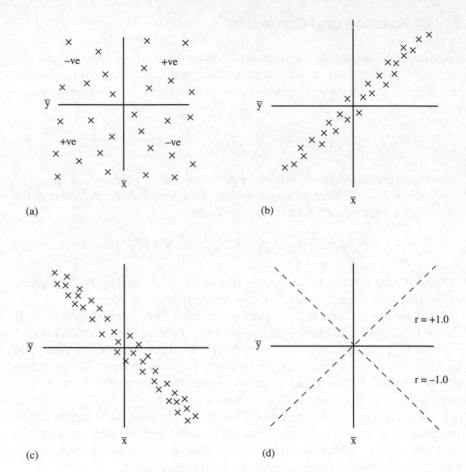

Figure 2.1 Covariance and correlation

In the case of Figure 2.1(a) the correlation coefficient will be zero. In Figure 2.1(b) and Figure 2.1(c) the correlation coefficient will have intermediate values within the range ± 1.0.

2.4 Normal Distribution

Symmetric bell-shaped distributions of the type shown in Chapter 1, Figure 1.2 can be modelled in statistical terms using the so-called *normal distribution* (sometimes referred to as the Gaussian distribution after the

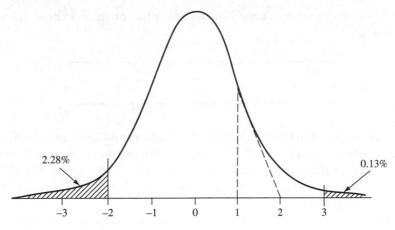

Figure 2.2   Normal distribution: $\bar{x} = 0$, $\sigma = 1.0$

celebrated German mathematician). The profile of this distribution is shown in Figure 2.2.

As seen in this diagram the mean is zero, the standard deviation is unity, and the tails of the distribution extend to three standard deviations (and beyond, to infinity).

The equation for the normal frequency curve is

$$\phi(x) = \frac{1}{\sqrt{2\pi}}\exp\left(-x^2/2\right)$$

The area to the left of the *ordinate* at x is given by

$$\Phi(x) = \frac{1}{\sqrt{2\pi}}\int_{-\infty}^{x}\exp\left(-t^2/2\right)dt$$

Extensive tables of the values of $\Phi(x)$ and $\varphi(x)$ are to be found elsewhere in the standard statistical literature, but it is the area to the right of the ordinate that is of special interest to production engineers because it can be used to quantify the proportion of rejects falling outside specification tolerance limits. Values of the variate x corresponding to specific percentages of outliers $P\%$ are tabulated for convenient reference in Appendix E.

In the following schematic diagram figures extracted from the table show that only a very small fraction (less than 0.2%) of the area under the normal distribution curve lies beyond plus or minus three standard deviations.

P (%)	0.2	0.1
X	2.88	3.09

In theory the distribution extends to plus/minus infinity, but beyond three standard deviations the height of the ordinate becomes vanishingly small. Within three standard deviations of the mean the normal distribution provides a good fit to many of the distributions encountered in engineering data.

In some quarters extrapolations are made beyond *three-sigma* and minutely small probabilities are quoted in parts-per-million, but the practice is dubious and millions of test results would be necessary to validate it. If an increased margin of safety is necessary it is more sensible to specify this in terms of the same scale of measurement as that which was used to record the data, or as a multiple of the standard deviation.

One of the commonest applications of the normal distribution in manufacturing engineering is to predict the proportion of units of product likely to fall outside specification limits. Consider the case of a product whose dimension is intended to meet a specification tolerance of 80 ± 0.3. The process is running slightly above target with a mean of 80.1 and a standard deviation of 0.14.

The tolerance limits can be expressed as multiples of the standard deviation (i.e. *standardised deviates*):

$$\text{Upper specification limit} = (80.3 - 80.1)/0.14 = +1.43$$
$$\text{Lower specification limit} = (79.7 - 80.1)/0.14 = -2.86$$

In the following schematic diagram these standardised deviates are inserted between adjacent values extracted from the table of percentage points of the normal distribution in Appendix E. The estimates $P \approx 7.5\%$ and $P \approx 0.2\%$ were arrived at by taking note that the value $x = 1.43$ is almost exactly midway between the ten and five *percentiles* and the value $x = 2.86$ is closer to the 0.2 percentile than to the 0.5 percentile.

P (%)	10.0	≈ 7.5	5.0	0.5	≈ 0.2	0.2
X	1.28	1.43	1.64	2.58	2.86	2.88

In this, and in subsequent schematic diagrams, bold type serves to focus readers' attention on the issue under discussion. The approximate equality sign $\approx$ identifies estimates that have to be determined by interpolating exact figures extracted from the tables. It is not suggested these schematic diagrams should be constructed on every occasion that reference is made to the tables in Appendix E. They are used here simply to demonstrate the process of visual interpolation.

From the above display it can be seen that the proportion of fall-out is as follows:

Above upper specification limit	$\approx 7.5\%$
Below lower specification limit	$\approx 0.2\%$
Total	$\approx 7.7\%$

If a full-dress table of the normal distribution function is used the precise estimate is 7.85%. Does the discrepancy of 0.15% really matter? If the process was brought back on target the tolerance limits would be at $x = \pm 0.3/0.14 = \pm 2.14$ standard deviations which is not quite midway between the two and one percentiles.

P (%)	2.0	≈ 1.6	1.0
X	2.05	**2.14**	2.33

The total fall-out would then be $2 \times 1.6 = 3.2\%$, less than half what it had been. This would be advantageous, but there would still be work to do to get the variability reduced. The source(s) of variability would have to be identified and brought under closer control. To eliminate fall-out the standard deviation would have to be reduced from 0.14 to 0.10 (one sixth of the overall tolerance). Even then, the process would have to be held strictly on target. If this was not possible a standard deviation less than 0.10 would allow some room for manoeuvre.

Before leaving the normal distribution it is worth noting that its standard deviation is not just a mathematical abstraction. Figure 2.2 shows that the point of inflexion at which the distribution curve changes from concave inwards to concave outwards occurs at one standard deviation and the tangent at that point intersects the base line at two standard deviations. In this way the standard deviation does provide a valid measure of the spread of the distribution, irrespective of the tails which extend to infinity in both directions.

2.5 Cumulative Frequency Distributions

Sometimes it is more convenient to look at a *cumulative frequency distribution* rather than a simple tabular distribution. The frequency distribution of dice scores used in the first chapter can be presented as a cumulative frequency distribution:

$R+Y+B$	Frequency	Cumulative (%)	Boundary
0, 1	0		
2, 3	1		
		1	3.5
4, 5	2		
		3	5.5
6, 7	7		
		10	7.5
8, 9	12		
		22	9.5
10, 11	15		
		37	11.5
12, 13	24		
		61	13.5
14, 15	17		
		78	15.5
16, 17	4		
		82	17.5
18, 19	9		
		91	19.5
20, 21	5		
		96	21.5
22, 23	3		
		99	23.5
24, 25	1		
26, 27	0		

The cumulative frequencies can be plotted on *probability graph paper* (see Figure 2.3) on which the vertical scale is symmetrical about the 50% mark, but the upper and lower graduations are 'stretched' towards infinity in such a way that the cumulative frequencies of a normal distribution will fall on a straight line.

It is apparent the points are close to a straight line so it is reasonable to attempt to fit a normal distribution to the data by drawing a line which lies between the points all along its length. This can be facilitated by placing a transparent straight edge on the paper, rotating it about its lower end until the points in the upper half of the graph are evenly divided, then rotating it

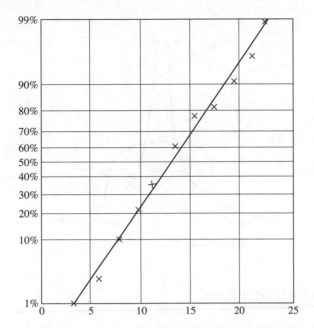

Figure 2.3 Cumulative probability distribution

about its upper end until the points in the lower half of the graph are evenly divided. Repeating this operation several times will provide a line which divides adjacent points more or less equally along the entire length.

The mean and standard deviation can then be read off on the horizontal scale. The median value is identified by the 50% graduation and because the normal distribution has perfect symmetry this is exactly equal to the mean. In the standard normal distribution the value of $\Phi(x)$ at $x = 1.0$ is 0.8413, so the standard deviation can be deduced from the 16% and 84% graduations.

This gives the values $\bar{x} = 13.0$ and $\sigma = 4.7$ which compare quite well with the computed values $\bar{x} = 12.92$ and $\sigma = 4.33$.

Probability graph paper can be used to examine the shape of distributions that do not conform to the normal pattern. Departure from normal which does not affect symmetry is referred to as *kurtosis*. Flat-topped distributions are described as *platykurtic*. 'Spiky' distributions are referred to as *leptokurtic* (Figure 2.4).

Logarithmic probability graph paper is available for dealing with asymmetric *skew* distributions (i.e. those with a long upper tail). These depend on the logarithmic transformation 'normalising' the data. Depending on the degree of skewness one-cycle, two-cycle, or three-cycle logarithmic probability graph paper may be used.

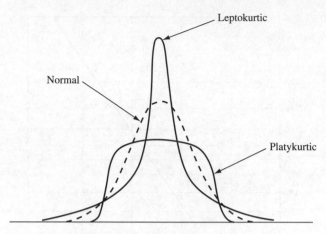

Figure 2.4 Distribution shapes

2.6 Binomial Distribution

The methods already described are suitable for data that can be expressed on a continuous scale of measurement, but on occasion engineers will encounter data that relates to events which either happen or do not happen, and the data simply records the number of such events in a given number of opportunities as a whole number. Several statistical distributions are available for handling such data. One is the so-called *binomial distribution*. Binomial = 'two names' (pass/fail, go/not-go, present/absent, active/passive, or whatever designation may be appropriate).

Given that the probability of an event happening is p, and the complementary probability of it not happening is $q = 1-p$, what is the probability of it happening r times out of n occasions? The answer is given by the terms of the binomial distribution

$$P(r) = \binom{n}{r}p^r q^{n-r}$$

where

$$\binom{n}{r} = \frac{n!}{r!(n-r)!}$$

To illustrate the principle let us consider an absurdly simple situation of a manufacturing process that is in dire trouble with a 30% failure rate and which is being monitored by applying a go/not-go test to samples of ten units of product drawn at random from the production line. On the average one would expect samples to contain three defective products, sometimes more, sometimes less, depending on random variation. What is the probability of getting a perfect sample, entirely free of defects? Are we ever likely to see a sample of ten defectives? The binomial distribution provides the answers:

R	$\binom{n}{r} = \dfrac{n!}{r!(n-r)!}$	$p^r q^{n-r}$	$P(r)$
0	1	0.0282475	0.03
1	10	0.0121061	0.12
2	45	0.0051883	0.23
3	120	0.0022224	0.27
4	210	0.0009530	0.20
5	252	0.0004084	0.10
6	210	0.0001750	0.04
7	120	0.0000750	0.01
8	45	0.0000321	0.00
9	10	0.0000138	0.00
10	1	0.0000059	0.00
			Total $= 1.00$

The calculation of

$$\binom{n}{r} = \frac{n!}{r!(n-r)!}$$

can be simplified by cunning manipulation of the factorials. For example, when $r = 4$

$$\binom{n}{r} = \frac{n!}{r!(n-r)!} = \frac{10!}{4! \times 6!} = \frac{10 \times 9 \times 8 \times 7 \times 6!}{4 \times 3 \times 2 \times 1 \times 6!} = \frac{630}{3} = 210$$

Alternatively, a device known as Pascal's triangle can be used. The individual values on each line are the sums of two adjacent values on the line above. The triangle is developed line by line until the requisite number of coefficients is available. In this case the eleventh line contains the required ten numbers.

```
                                        1
                                  1           1
                            1           2           1
                      1           3           3           1
                1           4           6           4           1
          1           5           10          10          5           1
    1           6           15          20          15          6           1
1     7           21          35          35          21          7           1
  8           28          56          70          56          28          8           1
9           36          84          126         126         84          36          9           1
10          45          120         210         252         210         120         45          10          1
```

The distribution of frequencies is represented graphically in Fig.2.5 (a).

Clearly, there is only a small probability of drawing a sample free from all defectives and it will not happen very often. The routine occurrence is for samples containing any number of defectives ranging from one to five. Two, three or four will be the most common. It is highly unlikely seven or more defectives will be found unless the process deteriorates still further.

The broad spread of this distribution is important to engineers. It shows that the control of a process using small samples of *attributes* data is insensitive. Whenever there is a choice control using variables data is preferable. Quite often there is no choice because of the nature of the defect. There is no measurable degree of defectiveness if a fuse blows, or if a short-circuit causes

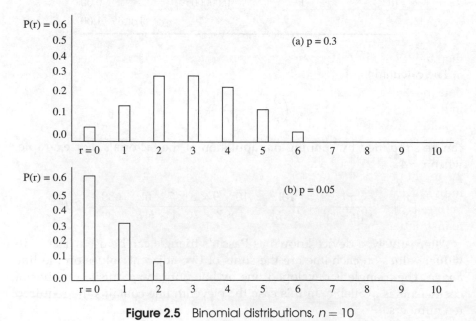

Figure 2.5 Binomial distributions, $n = 10$

apparatus to malfunction, or if a mechanical component fractures. In such circumstances the only remedy is to use a much larger sample and to make due allowance for the unavoidable variability between samples. But when a measured variable is available it is far better to use that data and not simply count the number of defectives falling outside a specified limit.

Referring again to Figure 2.5(a), it will be noted that the distribution is (roughly) symmetrical. That is not the case if the combination of n and p yields a very small mean value.

Consider what the distribution of probabilities would look like if the manufacturing process referred to above had been running at 5% failure rate instead of 30%

r	$\binom{n}{r} = \dfrac{n!}{r!(n-r)!}$	$p^r q^{n-r}$	$P(r)$
0	1	0.5987369	0.60
1	10	0.0315125	0.32
2	45	0.0016586	0.07
3	120	0.0000873	0.01
4	210	0.0000046	0.00
			Total $= 1.00$

The distribution of frequencies in Figure 2.5(b), is highly skewed. More than half the samples drawn from the production line would show no signs of defectiveness. Here again the lack of sensitivity of control by attributes data in small samples is revealed. We will return to distributions of this type in the next section when dealing with the Poisson distribution.

Before leaving the binomial distribution we will look at an application with a large data set. Figure 2.6, illustrates the problem of estimating the frequency of occasions on which a process operating at a mean defect rate of 2% might be expected to produce more than 25 defectives in a batch of 1000 individuals. The process of calculating the probabilities would be the same as before, except that there are many more possibilities ranging from just over 5 to just under 35 defectives per batch, with a mean of 20.

One approach to the problem would be to evaluate the eight individual terms for $r = 26, 27, \ldots 33$, but this would be tedious. Consider

$$P(26) = \frac{1000!}{26! \times 974!} \times (0.02)^{26} \times (0.98)^{974} \text{ etc.}$$

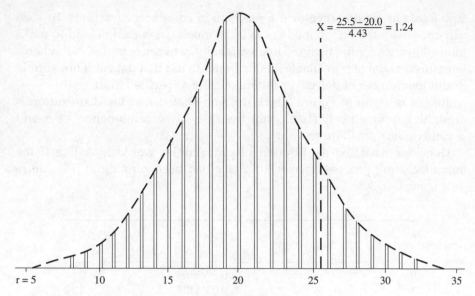

$$X = \frac{25.5 - 20.0}{4.43} = 1.24$$

Figure 2.6 Binomial distribution, $n = 1000$ $p = 0.02$

Fortunately there is a simple short cut. It can be shown mathematically that the envelope of the bar chart for a symmetrical binomial distribution conforms to a normal distribution with the same mean and standard deviation.

Mean of binomial $= np = 1000 \times 0.02 = 20.0$
Variance of binomial $= npq = 1000 \times 0.02 \times 0.98 = 19.6$
Standard deviation of binomial $\sqrt{npq} = 4.43$

To evaluate the total frequencies of the binomial distribution for the terms $r = 26$ and above we consider a notional ordinate of the normal distribution inserted midway between the binomial terms $r = 25$ and 26. The corresponding standardised deviate is

$$x = \frac{25.5 - 20.0}{4.43} = 1.24$$

Reference to the table of percentage points of the normal distribution in Appendix E will show that roughly 1 batch in every 10 would contain more than 25 defects. If for legalistic reasons a more precise estimate was needed reference to a full-dress table of the normal distribution would give the value of 10.75%.

2.7 Poisson Distribution

In mathematical terms the Poisson distribution is a limiting case of the binomial distribution in which small values of p are linked with large values of n. It has the advantage of being a single parameter ($\mu = np$) function which simplifies calculation. Analysing traffic data is one of the standard applications. It can also be used as an alternative to the binomial distribution provided $np < 1.0$.

The generating function is $P(r) = \dfrac{\mu^r}{r!} \times e^{-\mu}$ which produces a string of terms

$$e^{-\mu}\left(1, \mu, \frac{\mu^2}{2!}, \frac{\mu^3}{3!}, \frac{\mu^4}{4!}, \ldots\right)$$

or, more conveniently for calculation

$$P(0) = e^{-\mu}$$

$$P(1) = \mu \times P(0)$$

$$P(2) = \frac{\mu}{2} \times P(1)$$

$$P(3) = \frac{\mu}{3} \times P(2)$$

$$P(4) = \frac{\mu}{4} \times P(3)$$

Example: If the mean traffic rate is 3.2 vehicles per minute how often can we expect to see five or more vehicles per minute?

$$P(0) = e^{-\mu} = 0.04$$

$$P(1) = \mu \times P(0) = 3.2 \times P(0) = 0.13$$

$$P(2) = \frac{\mu}{2} \times P(1) = \frac{3.2}{2} \times P(1) = 0.21$$

$$P(3) = \frac{\mu}{3} \times P(2) = \frac{3.2}{3} \times P(2) = 0.22$$

$$P(4) = \frac{\mu}{4} \times P(3) = \frac{3.2}{4} \times P(3) = 0.18$$

$$P(5) = \frac{\mu}{5} \times P(4) = \frac{3.2}{5} \times P(4) = 0.11$$

$$P(6) = \frac{\mu}{6} \times P(5) = \frac{3.2}{6} \times P(5) = 0.06$$

$$P(7) = \frac{\mu}{7} \times P(6) = \frac{3.2}{7} \times P(6) = 0.03$$

$$P(8) = \frac{\mu}{8} \times P(7) = \frac{3.2}{8} \times P(7) = 0.01$$

$$P(9) = \frac{\mu}{9} \times P(8) = \frac{3.2}{9} \times P(8) = 0.00$$

The probability of five or more vehicles in one minute is the sum of $P(5)$, $P(6)$, $P(7)$, $P(8)$ i.e. $0.11 + 0.06 + 0.03 + 0.01 = 0.21$. We would expect to see five or more vehicles per minute on one occasion in five.

Now compare the Poisson distribution $\mu = np = 10 \times 0.05 = 0.5$ with the binomial $n = 10$, $p = 0.05$ in the previous section.

$$P(0) = e^{-0.5} = 0.607$$

$$P(1) = 0.5 \times P(0) = 0.304$$

$$P(2) = \frac{0.5}{2} \times P(1) = 0.076$$

$$P(3) = \frac{0.5}{3} \times P(2) = 0.013$$

	Poisson	Binomial	Discrepancy
$r = 0$	0.607	0.599	+0.008
$r = 1$	0.304	0.315	−0.011
$r = 2$	0.076	0.075	+0.001
$r = 3$	0.013	0.011	+0.002

It is at once apparent that the discrepancies are so trivial it is safe to simplify computation by using the Poisson distribution as a substitute for the binomial distribution when the mean value is small, $\mu < 1.0$.

2.8 Chi-squared Distribution

In engineering terms the statistical chi-squared function χ^2 may seem far removed from reality, but it has an important role to play in statistical analysis applied to practical situations. That is why the percentage points of the χ^2 distribution have been included in Appendix E.

χ^2 is defined as the sum of squares of a number of random normal deviates.

$$\chi^2 = u_1^2 + u_2^2 + u_3^2 + \ldots u_n^2$$

It is used when an observed set of data is being compared with the ideal of an 'expected' set of data that might exist in the same circumstances. In statistical terms it constitutes a *goodness of fit* test. The statistic is calculated using the formula

$$\chi^2 = \sum \frac{(O - E)^2}{E}$$

where O is an observed frequency and E is an expected frequency.

To use a simple example, the distribution of dice scores reported in Chapter 1 and displayed in Figure 1.1, might give rise to a suspicion that the thrower knowingly or inadvertently influenced the throw so as to minimise scores of 2 or 8. The following table shows which scores were contributing most to the value of χ^2.

Score	0	1	2	3	4	5	6	7	8	9
O	30	38	20	38	29	31	29	32	21	32
E	30	30	30	30	30	30	30	30	30	30
$O-E$	0	8	-10	8	-1	1	-1	2	-9	2
$(O-E)^2$	0	64	100	64	1	1	1	4	81	4

Hence

$$\chi^2 = \frac{1}{30} \sum (O - E)^2 = \frac{1}{30} (0 + 64 + 100 + 64 + 1 + 1 + 1 + 4 + 81 + 4)$$

$$= \frac{320}{30} = 10.7$$

How many degrees of freedom? 9, not 10. If random frequencies were allocated to any nine of the scores the tenth would be determined by the constraint that the total frequencies were three hundred. Reference to the table of percentage points in Appendix E shows that at $v = 9$ the 5% and 1% points are (approximately) at $\chi^2 = 17.0$ and $\chi^2 = 22.0$ respectively. The discrepancies between the scores are therefore not significant and must be regarded as purely due to chance.

Having used some trivial data to demonstrate the χ^2 principle we can now apply it to an important manufacturing situation with engineering connotations. A quality improvement campaign is about to be launched in a factory with three production lines on similar products. At the strategic planning stage data that has been collected over a period of time is being scrutinised. This purports to show the number of defects recorded on each line, classified in defect types A, B, C, D and E. There is much discussion on its interpretation. Some say the apparent differences between the performance of the three lines are purely due to random fluctuations that are an inevitable part of daily life in manufacturing industry. Others say the differences are highly significant and are due to raw materials, age of plant, skill of operators, building location, the weather, and anything else that comes to mind. There is a need for a sharper focus so that effort will not be wasted on pursuing unlikely causes.

The observed frequencies of occurrence O are given in Table 2.1. This also displays the total for each production line and for each category of defect as well as the grand total.

Table 2.2 shows the frequencies E that would be expected in each cell if there were no significant effects between defect categories or between production lines. These are arrived at by portioning the grand total into fractions

Table 2.1 Recorded defects (O)

Defects	A	B	C	D	E	Totals
Line 1	131	158	159	167	207	822
Line 2	212	190	179	200	189	970
Line 3	128	161	173	157	130	749
Totals	471	509	511	524	526	2541

Table 2.2 Expected Frequencies (E)

Defects	A	B	C	D	E	Totals
Line 1	152.4	164.7	165.3	169.5	170.2	822.0
Line 2	179.8	194.3	195.1	200.0	200.8	970.0
Line 3	138.8	150.0	150.6	154.5	155.0	749.0
Totals	471.0	509.0	511.0	524.0	526.0	2541.0

that are proportional both to the row totals and to the column totals. Thus, for the expected frequency of defect category A in production line 1:

$$E = \frac{822}{2541} \times \frac{471}{2541} \times 2541 = 152.4$$

The differences $O-E$ are recorded in Table 2.3. The contributions of individual differences to the total χ^2 calculated as $(O - E)^2/E$ and are entered in Table 2.4.

The total of all contributions gives a final value $\chi^2 = 28.447$. Before we can determine the significance of this value of χ^2 we have to consider, what are the degrees of freedom? If we were to allocate random values in the cells of the table while maintaining the existing row and column totals we could allocate up to four in the top row. The fifth value would then be determined by the other four in conjunction with the row total. The same would apply to the second row. That is as far as we could go, because all the values in the bottom row would now be determined by the two values in the first and second rows in conjunction with the column totals. The degrees of freedom are therefore $v = 4 \times 2 = 8$. This is consistent with the rule that in an $m \times n$ contingency table (to use statistical terminology) the degrees of freedom are $(m-1) \times (n-1)$. Reference to the table of percentage points of the χ^2 distribution in Appendix E shows that the critical values of χ^2 with $v = 8$ will be (approximately) 15.0 at 5% and 20.0 at 1%.

The evidence can now be summarised in Table 2.5 in which each item is listed in order of magnitude of its contribution to the total χ^2. Estimates of the

Table 2.3 Differences $(O-E)$

Defects	A	B	C	D	E
Line 1	−21.4	−6.7	−6.3	−2.5	36.8
Line 2	32.2	−4.3	−16.1	0.0	−11.8
Line 3	−10.8	11.0	22.4	2.5	−25.0

Table 2.4 Contributions to χ^2

Defects	A	B	C	D	E
Line 1	3.005	0.273	0.240	0.037	7.957
Line 2	5.767	0.095	1.329	0.000	0.693
Line 3	0.840	0.807	3.332	0.040	4.032

Table 2.5 Summary

Line	Defect	$(O-E)$	$\dfrac{(O-E)^2}{E}$	χ^2	
2	D	0.0	0.000	0.000	
1	D	−2.5	0.037	0.037	
3	D	2.5	0.040	0.077	
2	B	−4.3	0.095	0.172	
1	C	−6.3	0.240	0.412	
1	B	−6.7	0.273	0.685	
2	E	−11.8	0.693	1.378	
3	B	11.0	0.807	2.185	
3	A	−10.8	0.840	3.025	
2	C	−16.1	1.329	4.354	
1	A	−21.4	3.005	7.359	
3	C	22.4	3.332	10.691	
3	E	−25.0	4.032	14.723	$\chi^2 \approx 15.0$ at 5%
2	A	32.2	5.767	20.490	$\chi^2 \approx 20.0$ at 1%
1	E	36.8	7.957	28.447	

5% and 1% critical values are placed at the appropriate points alongside the table. This shows that the last two items $1/E$ and $2/A$ are significant and that a third item $3/E$ is on the borderline. The remaining items are not significant and their variability can be attributed to unidentified background sources common to the whole factory.

There are two distinct quality problems. The data in Table 2.1, records an average of 170 defects in each category on each production line across the whole factory. There is also a significant excess of category E defects on line 1, and of category A defects on line 2. In contrast, there is a significantly low number of category E defects on line 3.

The strategic plan for quality improvement should therefore develop two distinct lines of attack. The global problem of around 2,500 defects spread across the whole factory requires a search for a common source (or sources) that may exist either in the factory or elsewhere in raw materials or in product design. There should also be a search for local sources identified with the excess of defect A on line 2 and the excess of defect E on line 1. In the latter connection it would be useful to make a close comparison of line 1 with line 3 which has significantly fewer category E defects than either of the other two lines. It is possible that when the sources of the local problems have been identified they may also help to resolve the global problem.

Bibliography

Three early publications are still worth reading if they can be found in reference libraries. Shewhart (1931) is an all-time classic – the first-ever text book on statistical quality control. Tipett (1952) and Davis and Goldsmith (1972) identify work of fundamental importance in the textile and chemical industries. The latter was reprinted many times, starting with a first issue in the early 1940s and was widely used for several decades in tutorial courses for the application of statistics in industry:

Shewhart, W.A. 1931: *The Economic Control of Quality of Manufactured Product*. New York: Van Nostrand.

Tippet, L.H.C. 1952: *The Methods of Statistics*. London: Williams and Norgate.

Davies, O.L. and Goldsmith, P.L. (eds) 1972: *Statistical Methods in Research and Production with special reference to the chemical industry*. London and Edinburgh: Oliver and Boyd

A wide selection of modern texts on the subject of statistical methods in industry is available:

Bissell, D. 1994: *Statistical Methods for SPC and TQM*. London: Chapman and Hall.

Box, G.E.P., Hunter, W.G. and Hunter, J.S. 1978: *Statistics for Experimenters*. New York: John Wiley.

Chatfield, C. 1995: *Problem Solving – A Statistician's Guide*. London: Chapman & Hall.

Coleman, S., Greenfield, T., Jones, R., Morris, C and Puzey, I. 1996: *The Pocket Statistician*. London: Edward Arnold

Ford Motor Company. 1987: *Statistical Process Control Instruction Guide*. Brentwood, Essex.

Grant, E.L. and Leavenworth, R.S. 1988: *Statistical Quality Control*. New York: McGraw-Hill

Hines, W.W. and Montgomery, D.C. 1990: *Probability and Statistics in Engineering and Management Science*. New York, John Wiley.

Montgomery. D.C. 1991: *Introduction to Statistical Quality Control*. New York: John Wiley

Oakland, J.S. 1999: *Statistical Process Control*. London: Heinemann.

Owen, M. 1990: *SPC and Continuous Improvement*. Bedford, UK: IFS Publications.

Society of Motor Manufacturers and Traders. 1986: *Guidelines to Statistical Process Control*. London: SMMT

Wetherill, G.B. and Brown, D.W. 1991: *Statistical Process Control*. London: Chapman and Hall

Wheeler, D.J. and Chambers, D.S. 1992: *Understanding Statistical Process Control*. Knoxville: SPC Press.

Reference was made in the opening paragraph to an article setting out recommendations for a statistical engineering syllabus to be included in the engineering curriculum. This embraces both the managerial and the statistical aspects of quality:

Morrison, S.J. 1997: Statistical Engineering – the Key to Quality. *Engineering Science and Education Journal* **6**(3) 123–127 and *Engineering Management Journal* **7**(4) 193–198. London: Institution of Electrical Engineers.

Evidence of the importance of including a statistical element in the academic engineering curriculum is provided in a more recent article:

Morrison, S.J. 2002: The missing link. *Engineering Science and Education Journal* **11**(4) 133–138, London: Institution of Electrical Engineers.

3

Production

For the last 50 years (or longer) two standard statistical methods have been used to control the quality of manufactured products. These are; (i) *acceptance sampling schemes* in which decisions to accept or reject an entire batch of product are based on the inspection of relatively large samples; and (ii) *control charts* used to monitor manufacturing processes by taking small samples at regular intervals from the production line. Both techniques were American in origin arising from work in the telephone manufacturing industry. It is on record that Walter Shewhart introduced the control chart concept in a memorandum dated 1924 and subsequently Harold Dodge laid the foundations of sampling inspection using the Poisson distribution which was already familiar to telephone and telegraph engineers for analysing traffic on the networks.

Acceptance sampling inspection is applicable at the outgoing interface between manufacturer and customer and also at the incoming interface between supplier (of raw materials and components) and manufacturer. We will deal with sampling inspection first before going on to in-house applications of control charts which are the most obvious tool for attacking variability and establishing control on the production line. Other weapons which are well known to statisticians, though perhaps not so familiar to engineers, will be dealt with subsequently. These include significance tests, analysis of variance and linear regression.

Statistics for Engineers: an Introduction   S.J. Morrison
© 2009 John Wiley & Sons, Ltd

3.1 Sampling Inspection

Variability in a manufactured product can drive customers into the hands of competitors if they receive a proportion of goods that are outside agreed specification tolerance limits. Business lost in this way can be difficult to recover and is very damaging to profitability. Moreover, in a competitive market the product which displays least variability will always be more attractive than its competitors.

Customers receiving regular large consignments have long been accustomed to protect themselves by using lot-by-lot acceptance inspection sampling schemes in which the quality attributes of individual units of product are recorded as acceptable or rejectable and the decision to accept the lot, or to reject it and return it to the supplier, is based on the same findings. Given that samples of the same size are unlikely to contain exactly the same number of rejects on every occasion there are inherent risks of incorrect decisions both for the manufacturer and for the customer. These risks can be evaluated using the Poisson distribution which was introduced in Section 2.7.

In the context of sampling inspection, if the average number of rejects per sample is μ (not necessarily an integer) then the probability of a sample containing r rejects (r being an integer) is given by

$$P(r) = \frac{\mu^r}{r!}e^{-\mu}$$
$$= e^{-\mu}\left[1, \mu, \frac{\mu^2}{2!}, \frac{\mu^3}{3!}, \frac{\mu^4}{4!} \cdots\right]$$

Thus, if the average number of rejects per sample is 1.8 the Poisson probabilities will be:

Defects	Probabilities
0	$P(0) = e^{-1.8} = 0.1653$
1	$P(1) = P(0) \times 1.8 = 0.2975$
2	$P(2) = P(1) \times \dfrac{1.8}{2} = 0.2678$
3	$P(3) = P(2) \times \dfrac{1.8}{3} = 0.1607$

$$4 \qquad P(4) = P(3) \times \frac{1.8}{4} = 0.0723$$

$$5 \qquad P(5) = P(4) \times \frac{1.8}{5} = 0.0260$$

$$6 \qquad P(6) = P(5) \times \frac{1.8}{6} = 0.0078$$

$$7 \qquad P(7) = P(6) \times \frac{1.8}{7} = 0.0020$$

$$8 \qquad P(8) = P(7) \times \frac{1.8}{8} = 0.0005$$

$$9 \qquad P(9) = P(8) \times \frac{1.8}{9} = 0.0001$$

$$10 \qquad P(10) = P(9) \times \frac{1.8}{10} = 0.0000$$

This shows that the most likely occurrence will be samples containing either one or two defects, but samples containing none or samples containing three or four will be quite common. Samples containing five or more defects are a possibility, but will occur only very rarely.

For any given sampling plan specified in terms of sample size and accept/reject numbers (differing by unity) repeated Poisson calculations for trial values of μ enable the *operating characteristic* to be plotted. Typical characteristics are shown in Figures 3.1 and 3.2. These reveal the inherent weakness of sampling inspection.

Figure 3.1 is the operating characteristic of a sampling plan designed for an acceptable quality level (AQL) of 1.0%, but even at that level there is a small possibility of the lot being wrongly rejected (the *producer's risk*). Yet there is only an even chance of the lot being correctly rejected when it contains 3.33% defective.

Figure 3.2, is the operating characteristic of a 'zero acceptance' plan which might be used for critical defect categories. Although the AQL is stipulated to be 0.01% there will still be a one-in-ten chance of accepting lots in which one defective unit of product appears in every 500 (the *consumer's risk*).

Strictly speaking the correct distribution to use for analysing a sampling plan is the binomial distribution, but the formula for the Poisson distribution embracing only a single parameter is much simpler to use

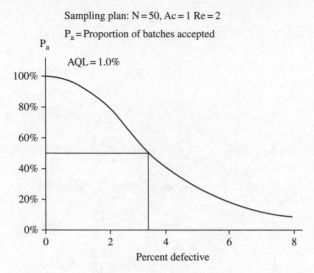

Figure 3.1   Operating characteristic for small sample

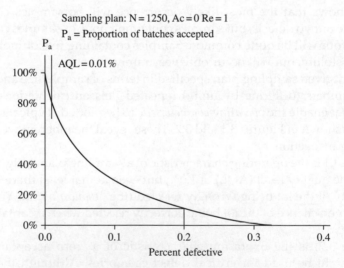

Figure 3.2   Operating characteristic for large sample

and the results are sufficiently accurate for practical purposes in the region of the AQL.

It is clear that although customers' acceptance sampling inspection maintains pressure towards quality improvement it does not provide sufficient protection to customers. Manufacturers who place too much reliance on sampling inspection expose themselves to the danger of dissatisfied customers looking elsewhere. Instead, they should concentrate on securing closer control of their own production processes and on tracking down sources of variability wherever they exist. The possibility of identifying variability that has been built into the product by decisions taken in design or in research and development should not be overlooked. More of that in subsequent chapters.

Finally, manufacturers should not forget they, too, are customers of their suppliers of raw materials and component parts the quality of which will support (or undermine) their own product. They should not rely only on incoming sampling inspection, but should make it a requirement that suppliers will provide continuing records of the quality of their materials and the steps taken to maintain their quality standards. By the same token the manufacturer should be willing to enhance the reputation of its products by providing a similar service to its customers.

3.2 Control Charts

The inherent weakness of acceptance sampling inspection underlines the importance of using control charts on the production line. Nowadays there is a great variety of control charts applicable in different situations, but the two basic types to be discussed here are the Shewhart chart which has been in general use for over 50 years and the *cusum* chart which has been around for at least 40.

Shewhart control charts are so well known it is hardly necessary to describe them here, yet some discussion of the underlying principles will not be out of place. Measurements are made on small samples (usually two, four or five individuals) drawn at intervals from a continuous flow on a production line. The sample means are plotted as a time series (Figure 3.3).

Control limits are ruled on the chart for the purpose of identifying abnormal behaviour of the process. How are these limits set? How are they to be interpreted? What is the underlying principle?

When successive samples of n individuals are drawn from a population with mean μ and standard deviation σ the distribution of sample means will tend towards normality. (This phenomenon was noted in Chapter 1 when discussing the totals of independent dice scores.) When statisticians refer to the standard deviation of a particular statistical

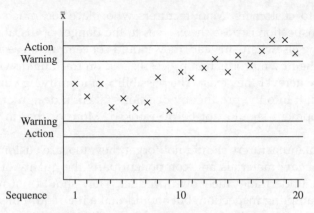

Figure 3.3 Control chart for sample mean

function they use the term *standard error*. By a relatively simple piece of mathematics it can be shown that the standard error (S.E.) of the mean of a sample of size n from a population with standard deviation σ is given by the relationship.

$$S.E.(\bar{x}) = \frac{\sigma}{\sqrt{n}}$$

It must be remembered that when control charts were first introduced hand-held calculators and desk-top computers were not even being dreamt about, so how could the standard deviation σ be determined from a sum of squares in a factory floor environment? Statisticians to the rescue! The *expectation of range* in samples from the normal distribution with $\mu = 0$ and standard deviation $\sigma = 1$ had been investigated with the following results:

Sample size	Range
2	1.1284
3	1.6926
4	2.0589
5	2.3261

This relationship was used in reverse so that the standard deviation could be estimated by multiplying the average sample range with a factor $1/d_n$ which is the reciprocal of the expectation of range. The average sample range

$\overline{w}$ could then be determined on the factory floor without much arithmetical difficulty. Thus:

$$s = \frac{1}{d_n} \times \overline{w}$$

It was the problem of factory floor arithmetic that made calculating the mean of samples of size $n = 3$ unpopular. Short cuts in mental arithmetic facilitated division by two, four or five, but not three. Typically, samples of two, four or five individuals were in common use.

Finally, the appropriate percentiles of the distribution of sample means were used to determine 1/40 *control limits* and 1/1000 *action limits* as shown in Figure 3.4.

All of the above statistical information was built into control limit factors, which are still published in handbooks of statistical quality control such as BS 5700.

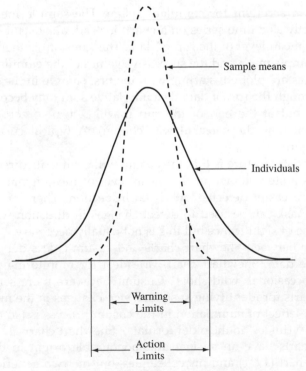

Figure 3.4 Control chart limits

It should be clearly understood that the function of a Shewhart control chart is to differentiate between short-term within-sample variability and long-term between-sample variability. In the absence of the latter not more than 1 point in 40 should lie outside the 1/40 warning limits. If the frequency of points outside the warning limits increases, and most certainly if points fall outside the 1/1000 action limits, corrective action should be taken to bring the process back on target.

Sample mean control charts provide a sensitive instrument for detecting and controlling between-sample variability arising from the process mean drifting away from its target, but if the within-sample variability is considered to be excessive we must dig deeper to find contributory causes. That leads directly to other statistical techniques which will be discussed later. But first, we must look at cusum charts.

3.3 Cusum Charts

'Cusum' is an acronym for cumulative sum. The sample means are not plotted directly as a time series on a chart. Instead, a constant is chosen at or about the mean level of the run of data. This constant is then subtracted from each sample mean and the successive values of the cumulative sum of the differences are plotted, starting with the first sample in the series. As it proceeds through the run of data the cumulative sum may become positive or negative, but at the end of the run it will return to zero (or nearly, depending on how the constant was chosen). A typical cusum chart is shown in Figure 3.5.

A characteristic feature is that the cusum will change its direction whenever there is a step change in the mean level of the original data. These *breakpoints* are easily detected by visual inspection. There are no control limits. A V-mask can be used to detect changes in direction which are on the borderline of significance but this is not usually necessary.

The choice between Shewhart charts and cusum charts depends on the nature of the data. Shewhart charts are ideal for monitoring progressive trends and occasional 'wild shots'. Cusum charts are more sensitive than Shewhart charts for identifying sustained step changes in the mean level of the data. The effect of summation in the cusum process is (so to speak) to 'iron out the wrinkles' in the older format of Shewhart charts.

Control charts have an ancient and honourable origin in the work of Walter Shewhart (1931) and his colleagues, but the two generic types discussed so far (sample mean and cusum) are by no means the only ones that

#	1	2	3	4	5	6	7	8	9	10
x	6.3	6.7	8.7	6.0	5.3	5.3	8.7	4.7	2.7	3.7
x–5.2	1.1	1.5	3.5	0.8	0.1	0.1	3.5	–0.5	–2.5	–1.5
Cusum	1.1	1.6	6.1	6.9	7.0	7.1	10.6	10.1	7.6	6.1
#	11	12	13	14	15	16	17	18	19	20
x	1.7	4.7	4.7	4.7	2.0	6.0	4.7	5.0	7.0	6.0
x–5.2	–3.5	–0.5	–0.5	–0.5	–3.2	0.8	–0.5	–0.2	1.8	0.8
Cusum	2.6	2.1	1.6	1.1	–2.1	–1.3	–1.8	–2.0	–0.2	0.6

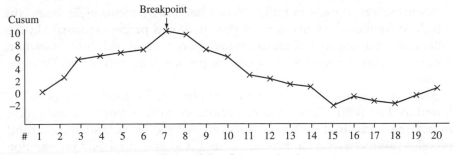

Figure 3.5   Cusum chart

modern engineers should be aware of. Control charts are still hot topics with statisticians and quality engineers. To quote a speaker at the Industrial Statistics in Action 2000 Conference (Cox 2001) 'The volume of papers still being published on control charts and quality control in general points to a continuing interest in the broad range of charting procedures'.

This view was borne out by the number of papers presented at the conference in which control charting appeared as a topic – ten in total. These are listed at the end of this chapter where they are numbered (i)–(x) for convenient reference here. Each one is of interest in the present context.

In (i) there is a report of a successful application of statistical methods for improving communication between departments in a soft drinks firm that has been modernising its management culture. A cusum chart identifying an unexplained change in mean fill level is one example.

Papers (ii), (iii) and (iv) all report on the application of statistical methods in the management of gas transportation in a complex national network with links to the Republic of Ireland and continental Europe. The relevant criteria were found to be quite different from those which apply in a typical manufacturing process. An interesting new development was the modification of the Shewhart chart by introducing customised 'review' limits at 90% inside traditional warning limits at 95% and action limits at 99.7%.

Next on the list, (v) is a theoretical examination of a generalised control chart procedure embracing Shewhart sample mean charts, cusum charts and exponentially weighted moving average charts. A new approach is demonstrated using spreadsheets and artificial neural networks which are computationally equivalent to how the brain is thought to work.

A demonstration is given in (vi) of new forms of exponentially weighted moving average charts in which the usual statistics are replaced by linear combinations of order statistics. These are shown to be advantageous when the distribution departs from normality.

A distinction is made in (vii) between parts manufacturing industry (in which conventional control charts originated) and process industry. In the latter case modified control limits are necessary to cope with autocorrelation between successive samples that creates problems in conventional sample mean charts.

Summed rank cusum charts which employ nonparametric methods are introduced in (viii). Three demonstrations of their application are given: measuring industrial effluent, assessing levels of potentially harmful proteins produced by an industrial process, and industrial land reclamation in the face of harmful waste materials. It is shown these are more suitable than conventional charts for monitoring environmental data.

Linear regression, in which a relationship between a predictor variable and a response variable is established, will be described in Section 3.6. The relationship may change with time. If it is used for controlling an industrial process it is important that such changes can be monitored. That is the purpose of the control charts described in (ix). One example dealing with crack growth in steel rails is of topical interest.

The last of this series of papers on control charts, (x) draws a distinction between using cusum charts for monitoring current processes and for retrospective (historical) data analysis. A practical recursive method is presented for breakpoint determination and significance assessment which can be automated in suitable software.

This series of ten papers demonstrates quite clearly that although traditional forms of sample mean charts and cusum charts are well established and continue in use there is scope for developing new forms of control charts to cope with unusual circumstances encountered in specific situations.

References to control charts are to be found elsewhere in the literature. The rights and wrongs of control charts have been discussed in a journal of the Royal Statistical Society (Caulcutt 1995). Control charts for monitoring two or more quality characteristics simultaneously are available (Jones 1995).

Control charts receive extensive coverage in SPC books (Bissell 1994, Oakland 1999, Ryan 2000, Wheeler and Chambers 1992). One book deals exclusively with cusum charts (Woodward and Goldsmith 1964).

3.4 Significance Tests

Sometimes there will be an apparent difference between measurements of nominally identical groups such as products from parallel production lines, or raw materials and components from separate suppliers. This leads into the rather tricky area of statistical significance testing. Significance tests all follow the same general pattern, beginning with a *null hypothesis* which asserts that the apparent effect does not exist. The data from independent sources is then lumped together and a suitable statistical formula is used to determine the probability that such a set of data could arise by chance. Depending on whether the probability is strong (say better than 5%) or weak (say less than 1.0%) the verdict may be 'not significant', 'significant' or 'highly significant' at a stated level of probability.

The statistician is looking at the suspect effect against the background of inherent variability. If the verdict is 'not significant' it does not necessarily mean there is no effect. It may be that it cannot be seen clearly until other sources of background variability have been cleared away.

Two significance tests that are relevant to statistical engineering are the *variance ratio F-test* and *Student's t*, the latter named after Gosset, the statistician at the Dublin brewery of Arthur Guinness who published his work under the pseudonym 'Student'. Both tests are used for comparing two independent samples.

To apply these tests it is necessary to calculate the sums and sums of squares and the means and variances for each set of data.

$$\bar{x}_1 = \frac{1}{n_1}\sum x_1 \quad \text{and} \quad V(x_1) = \frac{1}{n_1 - 1}\sum (x_1 - \bar{x}_1)^2$$

$$\bar{x}_2 = \frac{1}{n_2}\sum x_2 \quad \text{and} \quad V(x_2) = \frac{1}{n_2 - 1}\sum (x_2 - \bar{x}_2)^2$$

On the assumption there is no significant difference in the means and that $V(x_1)$ is the larger the variance ratio is

$$F = \frac{V(x_1)}{V(x_2)}$$

On the assumption the variances are not significantly different and that $\bar{x}_1$ is the larger the difference between the means is compared with the standard error of the difference in the following formulae:

$$t = \frac{(\bar{x}_1 - \bar{x}_2)}{s\sqrt{(1/n_1 + 1/n_2)}}$$

where

$$s^2 = \frac{\sum(x_1 - \bar{x}_1)^2 + \sum(x_2 - \bar{x}_2)^2}{(n_1 - 1) + (n_2 - 1)}$$

The degrees of freedom are $\nu = (n_1 - 1) + (n_2 - 1)$.

The values of F and t calculated in the above formulae can then be compared with the percentage points of the t-distribution and the F-distribution in extensive statistical tables to be found elsewhere in the statistical literature, but the abbreviated tables in Appendix E will be more convenient for most practical purposes in statistical engineering. They have the advantage of displaying the essential character of the t- and F-statistics in a compact format that could be carried around in one's pocket if not in one's head.

The table of percentage points of the t-distribution show that a large value of t is required in the smallest samples ($n = 2$, $\nu = 1$) and that the significance levels (5%, 1%) for small samples are a long way apart. This requirement diminishes rapidly as sample sizes increase.

A cursory examination of the table of percentage points of the F-distribution should begin at the bottom right hand corner. With $F = 1$ it is clear that any discrepancy of variability between two infinitely large samples would be regarded as significant. Moving diagonally across the table to the top left-hand corner it appears that for samples of the smallest size ($n_1 = n_2 = 2$, $\nu_1 = \nu_2 = 1$) a very large value of F is required for 5% significance and an enormous value for 1% significance. The top line right across the table shows that this is also true irrespective of the size of the sample with the larger variance (n_1, ν_1) if it is being compared with another sample of the smallest size ($n_2 = 2$, $\nu_2 = 1$). All the columns in the table show that the requirement for very large values of F diminishes rapidly as the size of the sample with the smaller variance (n_2, ν_2) increases. This effect levels out as the sample size increases.

With an appreciation of the 'shape' of these two statistics it should not be difficult for a statistical engineer to make reasonable visual interpolations in these tables. After all, it is not the precise value of the percentage point that is important. It is the correct value of the statistic derived from the available data when viewed against the general background of the tables in Appendix E which matters. That is what a statistical engineer should use to guide his judgement on whatever technical issue that is under consideration.

Interpolation of the abbreviated tables need not be a source of error. Guidance on visual interpolation will be given in examples in subsequent pages.

The application of significance testing can be demonstrated in a case study. A component which was thought to be a critical source of variability in a manufactured product was purchased from outside suppliers with a specified tolerance of $75.0^{+0.010}_{-0.000}$ on a particular dimension (i.e. plus ten thousandths, minus zero on the nominal). Two suppliers were being used. Measurements on samples from each (thousandths over the nominal dimension):

| Supplier A | 6.6 | 3.0 | 8.4 | 11.0 | 6.4 | 10.2 | 6.8 | 6.3 | 7.1 | 8.5 |
| Supplier B | 7.2 | 4.3 | 0.7 | 3.5 | 4.8 | 7.8 | −1.0 | 6.6 | −0.2 | 8.8 |

On a superficial examination of the data both suppliers appear to be having difficulty meeting the specification. Supplier A has two individuals over the upper limit. Supplier B has two individuals below the lower limit. On the basis of extreme values components from supplier B appear to be slightly more variable than those from supplier A. What can statistical analysis do to clarify the situation?

Preliminary arithmetic:

$$\Sigma x_A = 74.3 \quad \Sigma x_B = 42.5 \quad \Sigma x_A^2 = 597.71 \quad \Sigma x_B^2 = 288.99$$

$$\Sigma(x_A - \overline{x}_A)^2 = \Sigma x_A^2 - \frac{1}{n_A}(\Sigma x_A)^2 = 597.71 - \frac{1}{10} \times (74.3)^2 = 45.66$$

$$\Sigma(x_B - \overline{x}_B)^2 = \Sigma x_B^2 - \frac{1}{n_B}(\Sigma x_B)^2 = 288.99 - \frac{1}{10} \times (42.5)^2 = 108.37$$

First, the variances must be compared.

$$V(A) = \frac{1}{n_A - 1}\Sigma(x_A - \overline{x}_A)^2 = \frac{1}{9} \times 45.66 = 5.07$$

$$V(B) = \frac{1}{n_B - 1}\Sigma(x_B - \overline{x}_B)^2 = \frac{1}{9} \times 108.37 = 12.04$$

$$F = \frac{V(\mathrm{B})}{V(\mathrm{A})}$$

$$= \frac{12.04}{5.07}$$

$$= 2.37 \ \text{ with } v_\mathrm{A} = 9 \text{ and } v_\mathrm{B} = 9$$

On the face of it, components from supplier B do appear to be more variable than those from supplier A but is the evidence conclusive? The value of F, rounded up to $F = 2.4$, is displayed in a schematic diagram along with values extracted from the table of percentage points of the F-distribution in Appendix E and with the estimate $F \approx 3.2$ for $v_1 = 9$, $v_2 = 9$ which has to be close to the value for $v_1 = 10$, $v_2 = 10$.

v_1	5	9	10
$P\,(\%)$	5	5	5
$v_2 = 5$	$F = 5.1$		$F = 4.7$
$v_2 = 9$		$F = 2.4$ $F = 3.2$	
$v_2 = 10$	$F = 3.3$		$F = 3.0$

It is clear that the value $F = 2.4$ is not large enough to be regarded as significant at the 5% level. It could occur by chance much more frequently than 1 in 20 occasions. Against the background of variability in both sets of data the suspicion that components from supplier B are more variable than those from supplier A cannot be supported. The suspicion remains, but it would require a much larger set of data to resolve it.

We can now address the apparent difference in mean levels by applying the t-test.

$$\bar{x}_\mathrm{A} = \frac{1}{n_\mathrm{A}} \sum x_\mathrm{A} = 74.3/10 = 7.43 \quad \text{and} \quad \bar{x}_\mathrm{B} = \frac{1}{n_\mathrm{B}} \sum x_\mathrm{B} = 42.5/10 = 4.25$$

$$s^2 = \frac{\sum(x_\mathrm{A} - \bar{x}_\mathrm{A})^2 + \sum(x_\mathrm{B} - \bar{x}_\mathrm{B})^2}{(n_\mathrm{A} - 1) + (n_\mathrm{B} - 1)}$$

$$= \frac{45.66 + 108.37}{(10 - 1) + (10 - 1)}$$

$$= 8.5572$$

$$S = \sqrt{8.5572} = 2.93$$

$$t = \frac{(x_A - x_B)}{s\sqrt{(1/n_A + 1/n_B)}}$$

$$= \frac{(7.43 - 4.25)}{2.93\sqrt{1/10 + 1/10}}$$

$$= 2.43 \quad \text{with} \quad v = 18$$

This value, rounded down to $t = 2.4$, is displayed in a schematic diagram along with values extracted from the table of percentage points of the t-distribution in Appendix E and with the estimate $t \approx 2.9$ for $v = 18$ and $P = 1\%$.

	10	18	20
$P = 5\%$	$t = 2.2$		$t = 2.1$
$P \approx 2\%$		$t = 2.4$	
$P = 1\%$	$t = 3.2$	$t = 2.9$	$t = 2.9$

It is apparent, $t = 2.4$ is significant at around the 2% level and so the null hypothesis is defeated. It is beyond reasonable doubt that a significant difference in mean level exists between the two suppliers and that (whether by design or by accident) supplier B is operating nearer to the mid-point of the specification at 5.00 thousandths over the nominal dimension.

However, a standard deviation as large as $\sigma_B = 3.47$ (the square root of $V(B) = 12.04$) cannot sit comfortably within a tolerance of $+ 10/-0$ (in thousandths). If the mean is held on a mid-range target at 5.00 thousandths above the nominal dimension the specification limits can be expressed as multiples of the standard deviation, $\pm 5.00/3.47 = \pm 1.44$. This value of the *standardised deviate* can be inserted between adjacent values extracted from the table of percentage points of the normal distribution in Appendix E.

P (%)	10.0	7.5	5.0
X	1.28	1.44	1.64

7.5% will fall outside of each limit, making a total reject rate of 15%. If the mean is allowed to drift away from a mid-range target by 2.0 thousandths the fall-out on one limit will increase more rapidly than the reduction on the other as indicated below:

P (%)	20.0%	19.5	10.0	5.0	2.2	2.0
X	0.84	0.86	1.28	1.64	2.02	2.05

The standardised deviation of the upper specification limit will now be $(10.0-7.0)/3.47 = 0.86$ and the fall-out will be 19.5% (an increase of 12.0%. The standardised deviation of the lower specification limit will be $(7.0-0.0/3.47 = 2.02$ and the fall-out will be 2.2% (a reduction of 5.3%) the total reject rate will be $19.5 + 2.2 = 21.7\%$ (an increase of 6.7%). There will be a strong incentive on supplier B to keep the process on target.

At this stage a third supplier who has been taking variability seriously appears on the scene, confident their components can meet the specification. The following test results are offered as proof.

Supplier C:	6.7	2.8	6.6	3.5	4.9	5.9	5.7	6.4	4.1	4.5

$$\sum x_C = 51.1 \qquad \sum x_C^2 = 277.87$$

$$\sum (x_C - \bar{x}_C)^2 = \sum x_C^2 - \frac{1}{n_C} \left(\sum x_C \right)^2 = 277.87 - \frac{1}{10} \times (51.1)^2 = 16.75$$

$$V(C) = \frac{1}{n_C} \sum (x_C - \bar{x}_C)^2 = \frac{1}{9} \times 16.75 = 1.86$$

$$\sigma_C = \sqrt{1.86} = 1.36$$

This compares very favourably with $\sigma_B = 3.47$ for supplier B. For supplier C there will be a margin of $\pm [5.00-(3 \times 1.36)] = \pm 0.92$ for the process to run off target before serious problems arise with individuals outside specification.

The case study not only demonstrates the virtue of using statistical methods to sharpen up the technical judgements that are part of daily life in manufacturing industry, but also the value of quality improvements resulting from reducing variability in manufactured products. Supplier C will have found that reduced variability actually saves money by eliminating the costs of scrap and rework. If the manufacturer places future orders for components with supplier C the other two will have lost part of their livelihood and they will find it extremely difficult to recover from their loss.

3.5 Analysis of Variance

When there are more than two sources of variability to be compared the t-test is not relevant. Instead, the technique of *analysis of variance* (AoV) is appropriate. This is applicable to sets of data with a hierarchic structure that can be broken down into groups, subgroups, etc. The sum of squares for the whole data set can be broken down into elements which identify with different aspects of the data structure. There is no limit to the number of branches or levels in the organisation of the data. The only difficulty is that each individual data structure has its own algebraic identity for partitioning the sums of squares. In fact, whole books have been written covering an immense variety of situations in which AoV is applicable.

There is, however, a useful short-cut if the algebra is translated into plain English (Morrison 1981)

$$
\begin{pmatrix} \text{Sum of squares between} \\ \text{subordinate groups within} \\ \text{superior groups} \end{pmatrix} = \sum \left(\frac{(\text{Subordinate group total})^2}{\text{Number of individuals contributing}} \right)
$$

$$
- \sum \left(\frac{(\text{Superior group total})^2}{\text{Number of individuals contributing}} \right)
$$

both summations over the whole data set.

In essence, this conforms to the practice of subtracting a correction factor from a crude sum of squares to arrive at the adjusted sum of squares. It will be noted that if the individuals in the 'superior' data set are treated as subordinate groups (of one individual in each group) within the superior group (i.e. the whole data set) this formula is identical with that for the sum of squares used in the calculation of variance of the whole data set.

A simple two-level analysis of variance will now be demonstrated using the data in Table 3.1 in which values of tensile strength have been coded. The data was collected in order to test a rumour that there were differences between the products delivered from nominally identical plants owned by a single company. If such differences did exist and were allowed to continue unchecked they would contribute to the variability of the product in the market place. In essence, this is an extension of the comparison between two individual data sets using the t-test, to comparisons between three or more data sets using the F-test.

Table 3.1 Tensile strength of product

		$X = (\text{Strength} -60.0) \times 10$			
Plant A	22	24	0	1	4
Plant B	7	5	4	12	2
Plant C	−4	−5	−4	14	−9
Plant D	−11	−4	−12	−9	−7

Totals A: 51, B: 30, C: −8, D: −43, grand total: 30

Crude sums of squares:

Between plants	$[51^2 + 30^2 + (-8)^2 + (-43)^2]/5 = 1082.8$
Between individuals	$22^2 + 24^2 + \ldots + (-9)^2 + (-7)^2 = 2060.0$
Grand total	$30^2/20 = 45.0$

Adjusted sums of squares:

Between plants within data set	$1082.8-45.0 = 1037.8$
Between individuals within plants	$2060.0-1082.8 = 977.2$
Between individuals within data set	$2060.0-45.0 = 2015.0$

These values can now be entered in Table 3.2

Table 3.2 Analysis of variance of tensile strength

Source of variation	Sum of squares	Degrees of freedom	Mean square
Between plants within data set	1037.8	3	345.9
Between individuals within plants	977.2	16	61.1
Total	2015.0	19	−

The mean squares are the sums of squares divided by their respective degrees of freedom. Since the data comprises a relatively small sample of the population that could exist the divisor has to be $(n-1)$ rather than n. The total number of individuals in the data is 20 so the total number of degrees of freedom is 19. These have to be apportioned between the two lines in the tabular analysis as 3 for the four plants, and $4 \times (5-1) = 16$ for the individuals within the four plants.

Finally, the variability between the plants can be compared with the variability within the plants using the variance ratio to test the mean squares:

$$F = \frac{345.9}{61.6} = 5.66$$

with 3 and 16 degrees of freedom.

This value, rounded up to $F = 5.7$, is displayed in a schematic diagram along with values extracted from the table of percentage points of the F-distribution in Appendix E and with the estimate $F \approx 5+$ for $v_1 = 3$, $v_2 = 16$ and $P = 1\%$.

P (%)	$v_1 = 2$ I	$v_1 = 3$ I	$v_1 = 5$ I
$v_2 = 10$	$F = 7.6$		$F = 5.6$
$v_2 = 16$		$F \approx 5+$ $F = 5.7$	
$v_2 = 20$	$F = 5.9$		$F = 4.1$

The precise value of the estimate $F \approx 5+$ does not matter. What does matter is that the value $F = 5.66$ in the analysis of variance appears to be significant at (or about) the 1% level. We can conclude that variability between plants is significant by comparison with variability within plants.

What does this mean in engineering terms? If the quality of the finished product is affected by the tensile strength of the product and if the firm is concerned about the variability of its product in the market place (as it should be) then an investigation should be launched to identify the source of the differences between the plants. This may lie in any or all of a number of areas, especially if the plants are located in different geographic regions with different operating conditions and supplies of raw material. The service that statistical analysis has provided for these enquiries is to eliminate the danger of waste of time and resources on what might have been no more than an unfounded rumour.

The same general principles which have been demonstrated in this comparatively simple example can be applied in much more complex situations. If data had been available for two or more production lines in each plant a three-level analysis would have shown whether the principal source of variability lay between the plants or within the plants between the production lines.

3.6 Linear Regression

A common situation in manufacturing industry is the use of one variable to control another related variable. Very often the relationship between the two lacks precision because of the presence of other factors which are contributing a measure of random variability. This situation can be dealt with using the statistical technique of *linear regression*.

At the opening of Chapter 1 on the nature of variability steel wire and glass bottles were taken as examples of relatively simple everyday products which nevertheless had many complex sources of variability in manufacturing processes that were easy to describe. The same manufacturing processes provide examples of the use of a process variable to control a property of the finished product.

In the wire industry individual customers will specify differing tensile strengths over a very wide range of values. To satisfy these choices alternative methods of wire drawing are used, but the primary factor is the tensile strength of the heat treated rod before drawing. This in its turn depends on the carbon content of the ingot from which the rod was rolled. It is well known that a linear relationship exists between tensile strength and carbon content, but this has to be determined from data collected in the context of the particular heat treatment process in the individual wire mill.

In the manufacture of glass bottles molten glass flows continually through an adjustable orifice at the forehearth end of the melting tank. The molten glass is cut into 'gobs' by an automatic shear. These are dropped into moulds, spiked to make them hollow, and then blown up with compressed air to fill the moulds from which they are then ejected. The weight of glass in each gob determines the wall thickness, and therefore the volumetric capacity, of individual bottles. By adjusting the orifice and measuring the capacity and the weight of individual bottles a relationship between weight and capacity can be developed.

It is necessary to collect process records to determine the relationship between the property of the product over which control is being exercised and the process factor which is being used to accomplish this. The problem is demonstrated in the following data in which X represents the control factor and Y represents the dependent property.

$X = 0.2$	1.0	1.6	2.0	2.0	2.2	4.4	5.0	5.2	5.2
$Y = 33$	58	67	45	67	70	71	51	63	78
$X = 5.6$	6.2	6.6	7.0	7.4	7.8	8.2	8.2	8.6	9.8
$Y = 53$	74	64	93	89	64	84	101	107	98

Plotting the data on an X–Y field (Figure 3.6.) shows that there is a suggestion of a linear relationship, but there is also a degree of 'scatter' with individual points dispersed at random on either side of any line that might be drawn to identify the relationship. Where does one place a straight

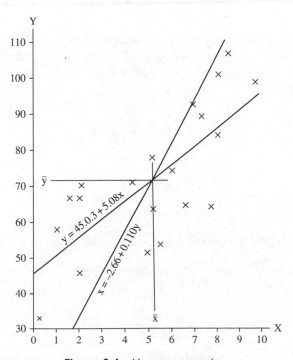

Figure 3.6 Linear regression

line through the points in order to get the most effective and reliable control over the whole range of operations?

We could begin by assuming that the values of X are known with absolute precision and that the 'error' is associated with Y. The statistical technique *regression of Y on X* will place a straight line through the centroid of the data $(\bar{x}, \bar{y})$ at an angle which will minimise the sum of squares of the y-deviates about the line. It is not surprising that this technique uses sums of squares and sums of products in the same way that they were used in an earlier chapter to calculate the correlation coefficient in bivariate data:

Preliminary arithmetic:

$$\Sigma x = 104.2 \quad \Sigma y = 1430 \quad \Sigma x^2 = 697.32 \quad \Sigma y^2 = 109408 \quad \Sigma xy = 8234.8$$

$$\Sigma(x - \bar{x})^2 = \Sigma x^2 - (\Sigma x)^2/n = 697.32 - \frac{1}{20} \times (104.2)^2 = 154.438$$

$$\Sigma(y - \bar{y})^2 = \Sigma y^2 - (\Sigma y)^2/n = 109408 - \frac{1}{20} \times (1430)^2 = 7163.0$$

$$\Sigma(x - \bar{x})(y - \bar{y}) = \Sigma xy - \frac{1}{n}\Sigma x \Sigma y = 8234.8 - \frac{1}{20} \times 104.2 \times 1430 = 784.50$$

Calculation of the correlation coefficient r is not part of regression analysis, but it is interesting to note that in the case of this data $r = 0.746$ confirms the impression that there is a moderately strong relationship between X and Y:

$$r = \frac{\Sigma(x_i - \bar{x})(y_i - \bar{y})}{\sqrt{\Sigma(x_i - \bar{x})^2 \Sigma(y_i - \bar{y})^2}}$$

$$= \frac{784.5}{\sqrt{154.438 \times 7163.0}}$$

$$= 0.746$$

Proceeding with regression analysis, the slope of the line is given by

$$b = \frac{\Sigma(x - \bar{x})(y - \bar{y})}{\Sigma(x - \bar{x})^2}$$

$$= \frac{784.50}{154.438}$$

$$= 5.08$$

The regression line must pass through the centroid of the data which is at

$$\bar{x} = \frac{1}{n}\Sigma x = \frac{104.2}{20} = 5.21$$

$$\bar{y} = \frac{1}{n}\Sigma y = \frac{1430}{20} = 71.5$$

The equation of the regression line can now be deduced as follows:

$$(y - \bar{y}) = b(x - \bar{x})$$
$$(y - 71.5) = 5.08(x - 5.21)$$
$$\therefore y = 45.03 + 5.08x$$

Hence, if it is intended that the process will deliver Y at 60 the control factor X should be set at

$$X = \frac{60.00 - 45.03}{5.08} = 2.95$$

The initial assumption, that all the 'error' in the data is associated with the variable Y could be questioned. If, instead, it is all assumed to be associated with the X variable there is a different result. The slope of the regression of X on Y is then given by

$$b = \frac{\Sigma(x - \bar{x})(y - \bar{y})}{\Sigma(y - \bar{y})^2}$$

$$= \frac{784.50}{7.163.0}$$

$$= 0.110$$

and the regression equation will be

$$x = -2.66 + 0.110y$$

Both lines are plotted on Figure 3.6; both pass through the centroid of the data, but there is an appreciable difference in slope which will affect the prediction at the extreme ends of the data. Given the nature of the assumptions it is probable that a true linear relationship between X and Y may lie somewhere between the two regression lines of Y on X and X on Y.

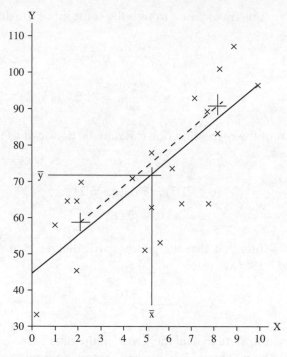

Figure 3.7 Three-group regression

There is, however, another form of regression analysis known as *three-group regression*, pioneered by Bartlett (1949) and evaluated by Gibson and Jowett (1957). This is concerned with fitting a linear relationship to two variates both subject to error. The method is demonstrated in Figure 3.7.

Quite simply, it is a matter of dividing the data into three equal groups, the lower, the middle and the upper. The centroids of the lower and upper groups are calculated in Table 3.3, and plotted in Figure 3.7.

A line is drawn linking the two outer centroids and then a line is drawn parallel to that through the centroid of the whole data. Given that there is no direct link between 'three-group' regression and 'least squares' regression it must be regarded as pure coincidence that the three-group regression of this particular set of data is almost identical with the least squares regression of Y and X with a slope of 5.17 (compared with 5.08). Three-group regression does not take account of the scatter of the data about the regression line, but it is simpler than least squares regression and it may appeal to engineers looking for a 'quick fix' on a functional relationship between two variables.

Table 3.3 Three-group regression analysis

Lower group		Upper group	
X	Y	X	Y
0.2	33	7.0	93
1.0	58	7.4	89
1.6	67	7.8	64
2.0	45	8.2	84
2.0	67	8.2	101
2.2	70	8.6	107
4.4	71	9.8	98
$\bar{x} = 1.91$	$\bar{y} = 58.7$	$\bar{x} = 8.14$	$\bar{y} = 90.9$

Bibliography

Statistical quality control (SQC) using control charts and other statistical methods originated in the first half of the twentieth century with the pioneering work of Walter Shewhart and his colleagues in America. In recent times there has been a revival of interest under the guise of statistical process control (SPC). Useful references for the content of this chapter are listed below.

In: Coleman, S.Y., Stewardson, D. and Fairbairn, L. 2000: *Industrial Statistics in Action*, University of Newcastle upon Tyne:

i) Bruce, D. and Coleman, S.Y. *Improving communication via quantitative management*. **I**, 304–311. (Also in *Journal of Applied Statistics*, 2001, **28** (3,4), 335–342, Carfax, Basingstoke.)

ii) Chambers, P.R.G., Piggot, J.L. and Coleman, S.Y. *SPC – A team effort for process control across four area control centres*. **I**, 74–92. (Also in *Journal of Applied Statistics*, 2001, **28**, (3,4), 307–324, Carfax, Basingstoke.)

iii) Coleman, S.Y., Gordon, A. and Chambers, P.R. *SPC – making it work for the gas transportation business*. **I**, 329–338. (Also in *Journal of Applied Statistics*, 2001, **28** (3,4), 343–351, Carfax, Basingstoke.)

iv) Coleman, S.Y., Arunakumar, G., Folvary, F. and Feltham, R. *SPC as a tool for creating a successful business measurement framework*. **II**, 243–253. (Also in *Journal of Applied Statistics*, 2001, **28** (3,4), 325–334, Carfax, Basingstoke.)

v) Cox. M.A.A. Towards the implementation of a universal control chart and estimation of its average run length using a spreadsheet: an artificial neural network is employed to model the parameters in a special case. **II**, 27–41. (Also in *Journal of Applied Statistics*, 2001, **28** (3,4), 353–364, Carfax, Basingstoke.)

vi) Elamire, E.A.H. and Seheult, A. *Control charts based on linear combinations of order statistics.* **II**, 264–279.

vii) Kaskavelis, E., Jonathon, P., Martin, E. and Morris, A.J. *Modification of the control limits for a multivariate autocorrelated process.* **I**, 232–243.

viii) Stewardson, D.J. and Coleman, S.Y. *Using the summed rank cusum for monitoring environmental data from industrial processes.* **I**, 121–137.

ix) Stewardson, D.J. Drewett, L., Silva, L., Mertens, J., Malone, R. and Bonta, J. *The use of control charts to monitor regression parameters: two practical applications.* **II**, 175–188.

x) Taylor, A.L., Tait, S.P., Porter, M.A., Perry, M.J. and Nicholson, R.W. *Automatic breakpoint detection for retrospective cumulative sum (Cusum) charts.* **II**, 198–210.

Other sources:

Bartlett, M.S. 1949: Fitting a Straight Line if both Variables are subject to Error. *Biometrics* **5**, 207.

Bissell, D. 1994: *Statistical Methods for SPC and TQM.* London: Chapman and Hall.

BS 5700 1984: *Process Control using Quality Control Chart Methods and Cusum Techniques.* London, British Standards Institute.

BS 5701 1980: *Number Defective Charts for Quality Control.* London, British Standards Institute.

BS 5703 1980–82: *Data Analysis and Quality Control using Cusum Techniques.* London, British Standards Institute.

Caulcutt, R. 1995: The rights and wrongs of control charts. *Journal of the Royal Statistical Society (C)* **44**, 3, 279–288. London: Royal Statistical Society.

Dodge, H.F. and Romig, H.G. 1959: *Sampling Inspection Tables: Single and Double Sampling.* New York: John Wiley.

Ford Motor Company. 1987: *Statistical Process Control Instruction Guide.* Brentwood, Essex.

Gibson, W.M. & Jowett, G.H. 1957: 'Three-Group' Regression Analysis. *Applied Statistics* **VI**, 2, 114–122, London: Royal Statistical Society.

Grant, E.L. and Leavenworth, R.S. 1988: *Statistical Quality Control.* New York: McGraw-Hill.

Ishikawa, K. (Transl. Loftus. J.H.) 1990: *Guide to Quality Control.* Chapman and Hall.

Jones, R. 1995: Control charts for monitoring two or more quality characteristics simultaneously. *Quality World Technical Supplement, September,* 119–126. London: Institute of Quality Assurance.

Juran, J.M. and Godfrey, A.B. 1999: *Juran's Quality Handbook,* 5th edn, Milwaukee, WI: American Society for Quality.

Montgomery, D.C. 1991: *Introduction to Statistical Quality Control.* New York: John Wiley.

Morrison, S.J. 1981: A note on the Teaching of Ao V. *The Statistician,* **30**, 4, 271–274. Bury St Edmunds: Institute of Statisticians.

Morrison, S.J. 1992: What's Wrong with Sampling Inspection? *Professional Engineering* **5**, 6, 16–17. London: Institution of Mechanical Engineers.

Oakland, J.S. 1999: *Statistical Process Control*. London: Butterworth-Heinemann.

Owen, M. 1990: *SPC and Continuous Improvement*. Bedford, UK: IFS Publications.

Ryan, T.P. 2000: *Statistical Methods for Quality Improvement*. Chichester: John Wiley.

Shewhart, W.A. 1931: *The Economic Control of Quality of Manufactured Product*. Van Nostrand.

Society of Motor Manufacturers and Traders. 1986: *Guidelines to Statistical Process Control*. London: SMMT.

Wetherill, G.B. & Brown, D.W. 1991: *Statistical Process Control*. London: Chapman and Hall.

Wheeler, D.J. and Chambers, D.S. 1992: *Understanding Statistical Process Control*. Knoxville: SPC Press.

Woodward, R.H. and Goldsmith, P.L. 1964: *Cumulative Sum Techniques*. Edinburgh: Oliver & Boyd for ICI.

4

Engineering Design

This chapter is an adaptation of the article 'Engineering Design: the Fount of Quality' published by IEE Publications in *Journal of Engineering Management* (Morrison 2000). *Variance synthesis* is a technique that is available to design engineers and which gives them the opportunity to forestall production problems and to enhance competitiveness of the product in the market place by avoiding the danger of inadvertently building excessive variability into the design.

4.1 Variance Synthesis

Over several decades the focus of statistical quality control has moved back progressively from the customer interface, through production, to research and product development. Variance synthesis carries the attack on product variability right back to the headwaters of engineering design with supremely attractive downstream benefits.

Engineering design can be regarded as an abstract intellectual process of synthesis – a putting together of notional elements of components and process factors with nominal values to achieve a desired standard of performance of a new process or product not yet existing. Usually (but not invariably) designs are based on mathematical formulae representing the technological relationships involved. These can be generalised in the format

$$X = f(x_1, x_2, x_3 \ldots x_n)$$

Statistics for Engineers: an Introduction   S.J. Morrison
© 2009 John Wiley & Sons, Ltd

where X is the target for some property of the product and x_1, x_2, etc. are parameters entered in the design formula.

These formulae are used to manipulate nominal values of design parameters such as dimensions of components and properties of materials, but they do not take account of variability which is at the heart of every quality problem in manufacturing industry. In the past variability was not investigated until its existence became manifest in the cost of scrap and rework or loss of business through customer dissatisfaction. More recently the attack on variability has been carried back through production into research and development using experimental methods (Antony and Kaye, 1996, Bendell 1988, Byrne and Taguchi 1986, Grove and Davis 1992, Montgomery 1984, Shainin and Shainin, 1988, Taguchi and Wu, 1985). However, experiments require the existence of a production line that is already running and they can be expensive and difficult to conduct.

There would be an obvious advantage in dealing with variability at the earliest conceptual stage of a new design and this is not impossible. A method for doing this was suggested nearly fifty years ago (Morrison 1957) and has since been updated (Morrison 1998, 2000). This uses the statistical measure of variability, $V(X)$, in the relationship

$$
V(X) \approx \left(\frac{\partial X}{\partial x_1} \right)^2 V(x_1)
$$
$$
+ \left(\frac{\partial X}{\partial x_2} \right)^2 V(x_2) + \ldots + \left(\frac{\partial X}{\partial x_i} \right)^2 V(x_i) + \ldots + \left(\frac{\partial X}{\partial x_n} \right)^2 V(x_n)
$$

The formula is exact for linear functions and is a good approximation for nonlinear functions if the standard deviation of each variable x_i is less than 20% of the mean. This condition is likely to be satisfied in most engineering situations.

By describing it as *variance synthesis* the method is associated with the synthesis of nominal values in conventional engineering design. By assembling the variances of contributory sources into a single variable it is the converse of the well-known statistical technique of *analysis of variance* described in Chapter 3, Section 3.5, in which a whole data set is broken down into its component parts.

The variance formula (sometimes referred to by statisticians as an error transmission equation) has an interesting structure. The partial derivatives $\partial X / \partial x_i$ govern the way in which the individual variances $V(x_i)$ contribute to the overall variance $V(X)$. Each partial derivative is a multiplier of the individual standard deviation σ_i and for that reason has to be squared in the variance formula. The partial derivative can be regarded as the slope of a gradient through which the individual variance $V(x_i)$ makes its contribution to $V(X)$.

If X is simply the sum of individual x_i values the partial derivatives are all unity and the variance of X is the sum of the individual variances. It is then obvious which are the dominant contributory sources of variability.

If the design formula is more complex than a straightforward sum the individual partial derivatives will differ from one another. Some will have steep gradients, others will be shallow. This will have a profound effect on determining the dominant contributory sources of variability, especially when the partial derivatives are squared. This phenomenon can not be deduced from technological considerations alone. It can only be perceived by looking at the situation through statistical eyes.

Two examples (one ancient, one modern) will be used to demonstrate the principle of variance synthesis. The first one comes from the early days of semiconductor devices. Before integrated circuits appeared on the scene individual components (diodes, transistors, etc.) were mounted as separate components on printed circuit boards. In the development of a metal-encapsulated transistor difficulty was experienced with miniature hermetic glass–metal seals used for passing conductors through the metal envelope. The problem was due to variability of volume of the tiny glass beads which were threaded over the conductors and then fused in eyeleted holes on a metal base. If there was too little glass the seal was incomplete. If there was too much the glass spread across the base and obstructed the mounting of the semiconductor crystal.

The beads were being made in the research laboratory by cutting short lengths off small diameter glass tubing on a relatively crude carborundum slitting wheel flushed with water. It was tempting to think the cutting process could be improved by purchasing an expensive tile cutting machine with a dynamically balanced spindle, a bonded diamond cutting wheel, a micrometer screw adjustment, and other refinements.

Before committing himself to action the research engineer visited the glass works to see the tubing being made. Glass tubing for the manufacture of fluorescent lamps was being made by the ton on a highly mechanised process, but the glass for miniature transistor seals was a special composition for matching the expansion characteristics of the metal base and the quantity required was much too small for mechanised production.

Instead, it was being melted in a small pot in an auxiliary furnace. A skilled craftsman would dip the end of a blowing iron into the pot, wind on a ball of molten glass, blow into the ball to make it hollow, roll it back and forth on a flat metal surface to make a hollow cylinder, all the while rotating it and swinging it above his head to prevent uneven wall thickness due to the effect of gravity. When the time was right his mate would stick a 'punty' on the opposite end of the cylinder and the two would walk backwards away from one another along the length of a tube drawing alley. A third would fan the molten glass with a board when he judged the size to be right.

There was little point in telling the craftsmen they would have to do better. They would simply have handed the blowing iron to the visitor saying, 'Show us how, Guv.' Instead, the engineer returned to his laboratory for a long hard think.

A sample of beads had their dimensions measured on a measuring microscope and their volume deduced from their weight measured on a sensitive torsion balance. A statistical summary of the measurements is shown in Table 4.1.

At first sight it appeared that length (having the largest variance) was the major problem, but developing the variance formula and inserting the data told a rather different story:

$$\text{Glass volume} = G = \frac{\pi}{4} (D^2 - B^2)L$$

where D = outside diameter, B = bore, L = cut length.

Hence

$$V(G) \approx \left(\frac{\partial G}{\partial D}\right)^2 V(D) + \left(\frac{\partial G}{\partial B}\right)^2 V(B) + \left(\frac{\partial G}{\partial L}\right)^2 V(L)$$

$$\approx \left(\frac{\pi DL}{2}\right)^2 V(D) + \left(\frac{\pi BL}{2}\right)^2 V(B) + \left(\frac{\pi}{4} [D^2 - B^2]\right)^2 V(L)$$

$$\approx \left(\frac{\pi \times 1.69 \times 1.92}{2}\right)^2 V(D) + \left(\frac{\pi \times 0.625 \times 1.92}{2}\right)^2 V(B)$$

$$+ \left(\frac{\pi}{4} [1.69^2 - 0.625^2]\right)^2 V(L)$$

$$\approx (5.09)^2 V(D) + (1.88)^2 V(B) + (1.94)^2 V(L)$$

$$\approx 25.95 \times V(D) + 3.55 \times V(B) + 3.75 \times V(L)$$

Table 4.1 Dimensions of glass beads

	Diameter	Bore	Length
Mean (mm)	1.69	0.625	1.92
Variance (mm$^2 \times 10^4$)	12.5	25.4	53.6

Hence, the contribution of each dimension to volume variance was as follows:

$$
\begin{aligned}
\text{Diameter} : 25.95 \times V(D) &= 25.95 \times 12.5 \div 10^4 = 0.0326(53\%) \\
\text{Bore} : 3.55 \times V(B) &= 3.55 \times 25.4 \div 10^4 = 0.0090(14\%) \\
\text{Length} : 3.75 \times V(L) &= 3.75 \times 53.6 \div 10^4 = 0.0201(33\%)
\end{aligned}
$$

Clearly, diameter rather than length was the culprit. The reason for this dramatic turnaround is shown in Figure 4.1.

The partial derivatives were different in magnitude:

$$
\frac{\partial G}{\partial D} = 5.09 \qquad \frac{\partial G}{\partial L} = 1.94
$$

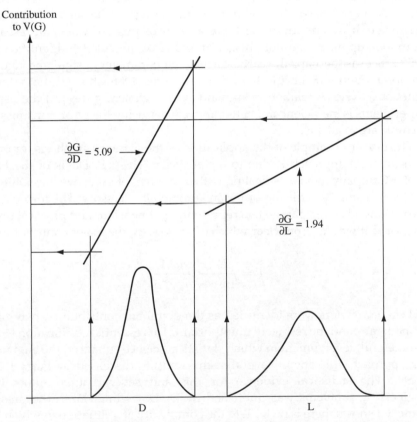

Figure 4.1 Effect of partial derivatives

Consequently, although cut length L had greater variability than tube diameter D its effect on volume G was diminished while that of D was strengthened.

The visit to the glass works had not been a waste of time because it was observed that the variability of tubing diameter was manifest as a gradual progressive taper, sometimes up, sometimes down, along a considerable length. Diameter could be brought under control by cutting the tubing into short lengths of 10 cm, gauging these, and classifying them into groups of consistent diameter. Each group would then be cut into beads of an appropriate length to give the correct volume. When this was done the production of miniature glass–metal seals went ahead successfully.

The thought process referred to earlier prompted the development engineer to adopt the role of amateur statistician by publishing an article in which he suggested the method might be used in engineering design (Morrison 1957). As happens so often with amateur gardeners the seed was slow to germinate. It was not until the 1980s and 1990s that an American research team took up his suggestion. Bisgaard and Ankenman (1995) identified the 1957 article as 'perhaps the earliest article on parameter design' and rated it 'important and much neglected'. Box and Fung (1986) showed that when a functional systems relation is known (as in engineering design) the use of experiments is inefficient and better approaches using the error transmission formula are available.

The modern example of the application of the variance synthesis (or error transmission) approach is due to a statistician, Dr T. P. Davis of the Ford Motor Company, who was dealing with a problem of excessive variability of torque in a small electric motor for an automobile accessory. The formula for torque embodied ten parameters representing dimensions and physical properties and it was far from clear which of these were the major contributors.

$$T = \frac{D^2 l_A l_M \theta \rho M}{l_D R_W [f_D - p_D]}$$

Two parameters were identified as the dominant contributors, two more as moderate, and the rest were unimportant. This essential clarification led to a successful development in which statistical science enhanced the engineering approach. It is one of several such examples discussed by Parry-Jones (1999). The statistical evidence for this and several other automotive engineering problems was published by Dr Davis in 2006. In the present context it shows there is no limit to the complexity of a situation in which the method of variance synthesis can be applied.

Both of the examples discussed above were in the context of research and development, but there does not appear to be any good reason why the method should not be used at the earliest conceptual stage of engineering design even before the first prototype has been made. Obstacles to its use are trivial and should be easily overcome. The application of differential calculus to design formulae should be within the competence of every professional engineer. It takes only minutes and hours, not days and weeks, to teach undergraduate engineers the principle of statistical variance as a measure of variability. The test data that provides nominal values in engineering reference tables could also provide the variance (or standard deviation) of the properties of materials. Modern instrumentation provides the opportunity to measure the variability of process factors.

The application of variance synthesis as a tool of engineering design has many attractions. Being derived directly from the design formula the variance equation is *technological* as well as statistical. It has *simplicity*, and therefore not likely to distract the design engineer from the main objective. It is *comprehensive*, embracing all the design parameters. It is *discriminatory*, identifying the dominant sources of variability at the outset. It is *instructive*, demonstrating in a unique way the build-up of contributory sources of variability. It is *inexpensive*, not requiring experiments or other costly investigations. It is *cost-effective*, producing tangible results with minimum cost. It is *upstream*, before any commitment is made to production. *Downstream benefits* include enhanced sensitivity of control routines as well as elimination of production problems identified in design. It is *strategic*, giving guidance to quality management planning. It has immense *academic research* potential, opening up the prospect of rewriting the entire cannon of engineering design in every branch of engineering.

Variance synthesis applied at the earliest possible engineering design stage of a new product should change the nature of experimentation in the development phase. By examining alternative tentative designs the design engineer has the opportunity to conduct exploratory virtual experiments at less expense than real experiments. By gaining an insight into sources of variability, factors which need to be tested in subsequent experiments in the laboratory or in the development workshop can be identified more clearly.

Finally, it has to be said that whether or not the design office can be persuaded to use variance synthesis it still remains a powerful tool in the hands of quality engineers for identifying the dominant sources of variability in products that are undergoing development or have already entered production.

4.2 Factors of Safety

Variance synthesis can help design engineers to ensure that a product will perform adequately in its intended environment with an adequate margin of safety, while at the same time not 'over-designing' by using too much energy or raw material, making the product too heavy or cumbersome in use, or too expensive to manufacture. Variability is a hazard in this situation, as in any other. Variance synthesis provides a statistical tool for calculating the margin of safety as well as for predicting performance. The procedure is represented in Figure 4.2.

X and Y are two variables which have to be balanced against one another. X might be the torque of a motive power unit required to overcome resistance Y in a driven mechanism, or X might be the strength of a structure required to bear a load Y. The variances of X and of Y can be determined independently by applying variance synthesis to each set of contributory factors.

As X and Y are independent the variance of the margin M will be the sum of the variances of X and Y. The critical value of M will therefore be three standard deviations of M. Anything less will lead to predictable disaster.

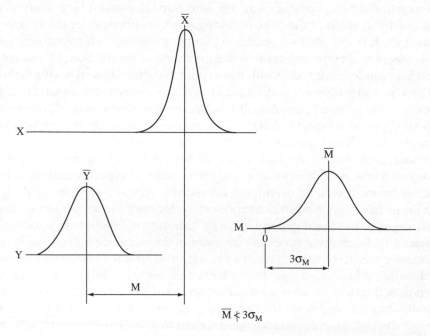

Figure 4.2 Statistical safety factor

From an engineering standpoint this is an eminently satisfactory procedure. It provides a margin of safety derived from the variability that is inherent both in the design and in the working environment. The safety margin is expressed in the same metric units as the design parameters and is therefore easy to understand. Engineering experience and judgement can be accommodated by adopting multiples of the standard deviation greater than three. If this is done the increased margin is proportional to the inherent variability and so the designer is not working in the dark. At the same time heavily over-designed situations are discouraged if the margin in a trial design is much greater than three standard deviations.

4.3 Tolerances

It has long been the custom for design engineers to write tight tolerances into specifications of products that they wish to see closely controlled. This can be disastrous if the tolerances are tighter than the process is capable of meeting.

Design engineers must recognise that products and processes are inseparable. If they wish to impose a tight specification on production they must satisfy themselves that the tolerance is realistic. Imposing an unrealistic tolerance simply creates a feedback loop in the production system which worsens control as a result of repeated adjustments being made to compensate for rejects on either side of the acceptance limits.

Designers who are unaware of certain basic statistical considerations can easily fall into the trap of setting unrealistic tolerances. In the development phase of a new product it is tempting to set tolerances on the basis of the range of measurements on a small number of prototypes, especially if the development engineer has taken a dislike to one or two individuals whose values were thought to be too high or too low to be consistent with their fellows.

The situation is depicted in Figure 4.3, which demonstrates the relationship between the standard deviation of a normal distribution and the expected values of the smallest and the largest values in a random sample.

It must be emphasised that the diagram is drawn accurately to scale and that it is not just an expression of opinion. The values depicted are based on sound mathematical reasoning and are borne out in practice. They are the values most likely to occur. On any one occasion the extreme values in a single sample may be larger or smaller, but the probability of occurrence decreases with distance from the most likely value.

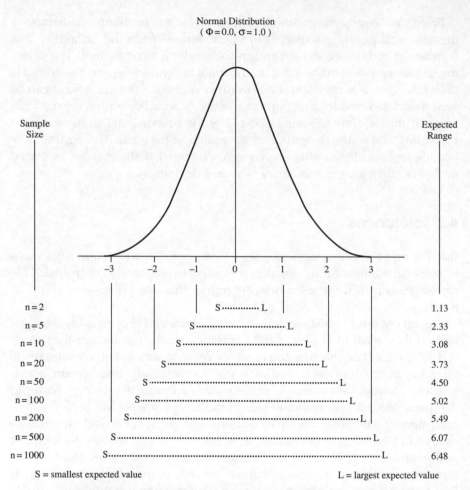

Figure 4.3   Engineering specification tolerance problem

The diagram shows that small sample data does not display the full extent of variability and that this may not be revealed until the full scale of mass production is encountered. The largest and smallest values in a small sample are very unlikely to come from the extremities of the parent distribution. If they do on any one occasion this will not happen again until a great many more samples are drawn.

Design engineers who write tight tolerances into a manufacturing specification after casually examining measurements on a small number of proto-types may be creating problems for production. The long tails of the distribution which are not apparent in small samples will inevitably appear in mass production. If these extend beyond the specified limits costly rejection and/or rework will be the inevitable consequence and in the long term there will be the prospect of loss of business leading to bankruptcy if competitors can do better.

Designers should always calculate the statistical variance of the prototype sample data using the '$n-1$' divisor discussed in Chapter 2, Section 2.2. Then, predicting the full extent of the variability to be expected in full-scale production, they should set tolerance limits not closer than three standard deviations on either side of the nominal.

Indeed, the limits should be set rather wider than three standard deviations so that any small drift of the process away from the ideal setting does not immediately incur a penalty. At the same time a requirement for the use of control charts on the final test parameters of the product should be written into the manufacturing specification so that the product will be maintained as close as possible to the nominal target. The manufacturing specification should also indicate which factors were identified in variance synthesis as the dominant sources of variability so that suitable control measures can be taken. Similar steps should be taken with regard to bought-in raw materials and components.

The old tradition of design offices setting tolerances without regard to statistical considerations and blaming production departments for inevitable nonconformance must be brought to an end. It was a contributory factor in the penetration of domestic markets by foreign manufacturers who were adopting an industrial culture linking statistical methods with technology and management. We were also losing international trade. There is no excuse for practising engineers neglecting statistical engineering nor for academic engineers leaving statistics out of the engineering curriculum.

4.4 The Future

If past mistakes are to be avoided engineering design must adopt new methods of coping with variability that is present in every engineering situation. Two trendsetting texts deserve attention.

Test Engineering – A Concise Guide to Cost-Effective Design, Development and Manufacture (O'Connor 2001) contains a chapter on Design Analysis. This deals with Quality Function Deployment (QFD), Computer Aided Design (CAD), Finite Element Analysis (FEA), Monte Carlo Simulation, Failure Modes and Effects Analysis (FMEA) and Fault Tree Analysis

(FTA). QFD and FMEA are particularly relevant to statistical engineering design. According to O'Connor the aim of QFD is to identify all of the quality requirements of a new product and to relate them to the product features that influence their achievement. FMEA is a method for tabulating all of the components (or functions) within a design and posing relevant questions about the probability of failure and the severity of effects. The book deals extensively with failure due to electrical and electronic stress as well as mechanical stress and there is an interesting chapter on software errors.

Designing Capable and Reliable Products (Booker *et al.* 2001) leans more towards mechanical engineering design and goes deeper into statistical methods. A chapter on Designing Reliable Products provides a detailed critical examination of statistical methods for probabilistic design. Twelve appendices occupying over one hundred pages at the end of the text contain a remarkable combination of statistical and technical information.

Both these texts are packed with illustrative examples. They constitute unique sources of information and reference data necessary for the pursuit of statistical engineering design.

Bibliography

Antony, J. and Kaye, M. 1996: Optimisation of core tube life using Taguchi experimental design methodology. *Quality World Technical Supplement*, March, 42–50. London: Institute of Quality Assurance.

Bendell, T. 1988: Taguchi comes to Europe. *Professional Engineering* **1**(4), 80–81. London: Institution of Mechanical Engineers.

Bisgaard, S. and Ankenman, B. 1995: Analytic Parameter Design. *Quality Engineering* **8**(1), 75–91. New York: Marcel Dekker

Booker, J.D., Raines, M. and Swift, K.G. 2001: *Designing Capable and Reliable Products*. *London*: Butterworth-Heinemann.

Box, G.E.P. and Fung, C.A. 1986: *Studies in Quality Improvement: Minimising Transmitted Variation by Parameter Design*. Report No.8, Center for Quality and Productivity Improvement, Madison, WI.

Byrne, D.M. and Taguchi, S. 1986: The Taguchi approach to parameter design. In: *Proceedings of the American Society for Quality Control 40th Anniversary Quality Congress*. Anaheim, CA, USA, May 19–21, 1986, 168–177.

Davis, D.P. 2006: Science, engineering and statistics. *Applied Stochastic Models in Business and Industry*, **22**, 401–430. Chichester: John Wiley.

Grove, D.M. and Davis, T.P. 1992: *Engineering Quality and Experimental Design*. London: Longman.

Montgomery, D.C. 1984: *Design and Analysis of Experiments*. New York: John Wiley.

Morrison, S.J. 1957: The Study of Variability in Engineering Design. *Applied Statistics* **VI**(2), 133–138. London: Royal Statistical Society.

Morrison, S.J. 1998: Variance Synthesis Revisited. *Quality Engineering* **11**(1), 149–155. New York: Marcel Dekker.

Morrison, S.J. 2000: Engineering Design – The Fount of Quality. *Engineering Management Journal* **10**(4), 194–196. London: Institution of Electrical Engineers.

O'Connor, P.D.T. 2001: *Test Engineering – A Concise Guide to Cost-effective Design, Development and Manufacture*. Chichester: John Wiley.

Parry-Jones R. 1999: Engineering for Corporate Success in the New Millennium, The 1999 Engineering Manufacturing Lecture. London: Royal Academy of Engineering.

Shainin, D. and Shainin, P. 1988: Better than Taguchi orthogonal tables. *Quality and Reliability Engineering International* **4**, 143–149.

Steiner, S.H. & Mackay, R.J. 2005: Statistical Engineering: *An Algorithm for Reducing Variation in Manufacturing Processes*. ASQ Quality Press, Milwaukee, U.S.A.

Taguchi, G. and Wu, Y. 1985: *Introduction to Off-Line Quality Control*. Nagoya: Central Japan Quality Control Association.

5

Research and Development

When the design and specification for a new product are ready it is customary to run preproduction trials to test the product and to detect problems that may not have been anticipated. Moreover, if the new product is not just an extension of an existing range, but is breaking new ground, it may be necessary to research unfamiliar technology. Such work comes under the colloquial heading 'R&D' (research and development).

The statistical techniques already discussed in Chapter 3 (Production) and Chapter 4 (Engineering design) are just as relevant to R&D operations. Three more will now be introduced for the R&D phase, namely *design of experiments, evolutionary operation,* and *multiple regression.*

All three are concerned with disentangling complex relationships in multivariate situations, but they do so in a different way. The statistical design of experiments is applicable when an experimental operation is being planned to reveal the effects of deliberate changes in selected factors. Evolutionary operation involves collecting data during a normal production run in order to seek a better combination of process factors. Multiple regression is applicable when data has been collected during normal production operations that embrace substantial changes in process factors so that their effects can be studied. The chapter ends with a brief review of other statistical techniques that are applicable in R&D, followed by a list of references.

Statistics for Engineers: an Introduction   S.J. Morrison
© 2009 John Wiley & Sons, Ltd

5.1 Design of Experiments

The statistical design of experiments owes its origin to the pioneering work of Fisher and Yates in agricultural research where it was normal to divide a plot of land into subplots and strips for the purpose of testing varieties of grain, different fertilisers, alternative methods of cultivation, etc. The underlying principle was that each strip of land would be treated to combinations of all factors so that each trial would contribute to the analysis of every factor being researched. The application of this principle made it possible to observe the way in which different factors interacted with one another as well as to study their direct effects.

This principle has much to commend it in the context of statistical engineering. Statistically designed experiments are more powerful and more efficient than traditional 'one-thing-at-a-time' experiments which in any case can never reveal interactions that are often very important.

In the early days of agricultural research the analysis of variance of experimental data was employed, but in recent times a graphical method termed *normal plots* has been introduced. Both methods are in current use and both were in evidence at the Industrial Statistics in Action 2000 conference (Coleman et al. 2000). Both will be described in this chapter.

There is virtue in looking at the numerical analysis approach first, to grasp some of the problems and principles of the statistical design of experiments before moving on to the graphical method of analysis which will be demonstrated alongside the analysis of variance of a moderately large four factor design. The whole field of experimental design is a very large one, and only a brief introduction will be attempted here.

First, some definitions appertaining to the statistical design of experiments:

A *designed experiment* is a planned set of trials or observations on a subject which is under investigation.

A *factor* or *treatment* is something which is varied at will (such as pressure, temperature, variety or source of material, etc.) in order to study the response of the subject. Factors may be *qualitative* or *quantitative*.

The *level* of a factor is one of a number of values or varieties chosen for the experiment.

A *treatment combination* is the combination of levels of every factor selected for a specific trial.

The *response* is the numerical result of a trial at a particular treatment combination which is measured on a parameter or characteristic of the subject.

An *effect* is the difference between responses at different levels of a factor, averaged over the whole experiment. A *main effect* is the effect of a single factor independent of all others. An *interaction effect* arises when the effect of one factor depends on the level(s) of other factor(s).

Replicate. An experiment comprising one trial at each treatment combination is a single replicate of the experiment. Hence there may be *multiple* or *fractional* replicates.

A *randomised* experiment is one in which the sequence of trials is randomised.

A *block* is a subset of trials within an experiment carried out on part of the experimental material likely to be more homogeneous than the whole.

Because of different background constraints there are likely to be differences in the most suitable designs for agricultural experiments and for engineering experiments, yet the statistical principles are common to both. The practice of dividing an agricultural field into plots to compensate for fertility trends across a field and to subdivide these plots into strips to accommodate numerous treatment combinations is straightforward. Engineering products and processes are less susceptible to subdivision in this way.

It is the results of agricultural experiments that are important rather than the trial crops themselves. Engineering experiments on a production line can be costly because they occupy valuable production time and the products of some trials are likely to be unsaleable. Nevertheless, there is much that can be learned by engineers from the efficiency and effectiveness of the statistical principles which were developed in designing agricultural experiments and which were subsequently employed successfully in the chemical industry.

Statisticians are often at pains to argue that statistical designs are more efficient and more effective than traditional 'one-factor-at-a-time' engineering experiments. There is merit in their argument, though there are difficulties to be encountered and avoided. The following simple example of the design, analysis and interpretation of a two-factor experiment in four trials will illuminate both sides of the argument.

Table 5.1, gives the responses of trials involving all four treatment combinations of two factors, each at two levels A_1 and A_2 with B_1 and B_2. Table 5.2 shows the responses coded by subtracting 50.0 and eliminating the decimal point to simplify the arithmetic. This table includes row and column totals that will be required for the arithmetic. The analysis of variance is carried out using the 'superior/subordinate group' formula given in Chapter 3, Section 3.5. The 'superior' group is the whole data set. The 'subordinate'

Table 5.1 Two factors without replication

	A_1	A_2
B_1	56.2	67.1
B_2	53.6	63.1

Table 5.2 Coded responses

	A_1	A_2	Totals
B_1	62	171	233
B_2	36	131	167
Totals	98	302	400

groups for factor A are the two columns and the 'subordinate' groups for factor B are the two rows. The 'subordinate' groups for the total sum of squares are the four individual responses.

Crude sums of squares

Factor A $(98^2 + 302^2)/2 = 100808/2 = 50404$
Factor B $(233^2 + 167^2)/2 = 82178/2 = 41089$
Individuals $(62^2 + 171^2 + 36^2 + 131^2) = 51542$
Correction factor $400^2/4 = 40000$

Adjusted sums of squares

Factor A $50404 - 40000 = 10400$
Factor B $41089 - 40000 = 1089$
Total $51542 - 40000 = 11542$

The sums of squares are now entered in the tabular analysis of variance (Table 5.3). The residual sum of squares is arrived at by subtracting the other two from the total. It is attributed to random variability from unidentified sources extending across the whole of the experiment. There are only three degrees of freedom (one less than the number of trials) so one must be assigned to each of the two factors and to the residual.

Table 5.3 Analysis of variance

Source of variation	Sum of squares	Degrees of freedom	Mean square
Factor A	10400	1	10400
Factor B	1089	1	1089
Residual	53	1	53
Total	11542	3	

The mean squares for factors A and B, arrived at by dividing the sums of squares by the degrees of freedom, can be compared with the residual mean square using the variance ratio F. For factor A

$$F = \frac{10400}{53} = 196.2$$

for factor B

$$F = \frac{1089}{53} = 20.5$$

These values are displayed in a schematic diagram for comparison with values extracted from the table of percentage points of the F-distribution for $v_1 = v_2 = 1$ in Appendix E.

$P >> 5\%$	5%	$P < 5\%$	1.00%
$F = 20.5$	$F = 161.4$	$F = 196.2$	$F = 4052$

It is apparent that the variance ratio $F = 196.2$ for factor A is significant at the 5% level. The variance ratio $F = 20.5$ for factor B is not. In engineering terms it can be accepted that factor A does have a real effect and the chance of this conclusion being wrong is less than 1 in 20. The effect of changing from A_1 to A_2 is $1/2\,[(67.1 + 63.1) - (56.2 + 53.6)] = 10.2$ units when averaged over the whole experiment and reverting to the original uncoded data. Factor B cannot be dismissed out of hand. There is a possibility that the lack of statistical significance may be due to a weakness in the design of the experiment arising from the exceptionally large values on the top line of the table of percentage points of the F-distribution in Appendix E.

The *efficiency* of this experiment is clear. If each factor had been tested in a separate experiment the same number of trials (four) would have been necessary. Depending on which pair of trials had been assigned to factor A the effect would have been recorded either as $67.1-56.2 = 10.9$ or as $63.1-53.6 = 9.5$. By taking the average over all four trials in the factorial experiment a more accurate estimate, 10.2, has been obtained. At the same time the possibility of a wildly inaccurate result due to an error in a single trial has been avoided.

The *effectiveness* of this experiment can be criticised. Factor B has been dismissed as not statistically significant, yet prior technical knowledge gave good reason to believe that it would have an effect, though not as powerful as factor A. One weakness of the design rests in the very high values of the variance ratio required for significance when only one degree of freedom is available for the denominator of the variance ratio. This can be seen by comparing the top line with subsequent lines in the table of percentage points of the F-distribution in Appendix E. It is clear that to have only one degree of freedom v_2 in the denominator variance is a situation to be avoided like the plague. Since the denominator is associated with the residual, designs which assign more than one degree of freedom to the residual are preferable.

One solution is to enlarge the scale of the experiment by using replication. There may, of course, be other reasons for enlarging the experiment in this way but in any event it gives us an escape from the 'top line'. Table 5.4 gives the coded responses for an extension of the above experiment in which each trial was replicated twice, making eight trials in all. The experiment now had seven degrees of freedom and it also opened up the possibility of looking for an interaction between the factors.

The row and column totals for the A and B responses are shown in Table 5.4. To study a possible A $\times$ B interaction it is also necessary to look at the total response within each cell created by the row and column intersections (Table 5.5).

Table 5.4 Coded responses, two factors with replication

	A_1	A_2	Totals
B_1	62, 55	171, 174	462
B_2	36, 43	131, 129	339
Totals	196	605	801

Table 5.5 A × B interaction cell totals

	A₁	A₂
B₁	117	345
B₂	79	260

The analysis of variance proceeds in the same way as before treating the rows and the columns as subordinate groups and the whole data set as the superior group. Similarly the cells formed by the row and column intersections are treated as a subordinate group, but because the variability between cells is enhanced by the variability between rows and between columns it is only the excess of the sum of squares for cells over the combination of the row and column sums of squares that can be attributed to an interaction.

Crude sums of squares

Factor A $(196^2 + 605^2)/4 = 404441/4 = 101110.250$

Factor B $(462^2 + 339^2)/4 = 328365/4 = 82091.250$

Interaction A × B: $(117^2 + 345^2 + 79^2 + 260^2)/2 = 206555/2$
$$= 103277.500$$

Individuals: $(62^2 + 55^2 + 171^2 + 174^2 + 36^2 + 43^2 + 131^2 + 129^2)$
$$= 103333.000$$

Correction factor $801^2/8 = 80200.125$

Adjusted sums of squares

Factor A $101110.250 - 80200.125 = 20910.125$

Factor B $82091.250 - 80200.125 = 1891.125$

Interaction A × B $(103277.500 - 80200.125)$
$$-(20910.125 + 1891.125) = 276.125$$

Total $103333.000 - 80200.125 = 23132.875$

The sums of squares are now entered in the tabular analysis of variance (Table 5.6). As before, the residual sum of squares is arrived at by subtracting the other three from the total. It is attributed to random variability from

Table 5.6 Analysis of variance

Source of variation	Sum of squares	Degrees of freedom	Mean square
Factor A	20910.125	1	20910.125
Factor B	1891.125	1	1891.125
Interaction A $\times$ B	276.125	1	276.125
Residual	55.500	4	13.875
Total	23132.875	7	

unidentified sources extending across the whole of the experiment. In eight trials there are seven degrees of freedom. One must be assigned to each of the two factors and to the interaction leaving four for the residual.

The first step of the analysis is to compare the interaction with the residual using the variance ratio:

$$F = \frac{276.125}{13.88} = 19.89 \text{ with } v_1 = 1 \text{ and } v_2 = 4$$

This value, rounded up to $F = 19.9$, is displayed in a schematic diagram for comparison with values extracted from the table of percentage points of the F-distribution in Appendix E. The estimate $F \approx 20$ for $v_2 = 4$ is based on the assumption it will be rather larger than for $v_2 = 5$, but nowhere near as large as for $v_2 = 2$.

v_1	1	1
$P(\%)$	5	1
$v_2 = 2$	$F = 18.5$	$F = 98.5$
$v_2 = 4$	$F \approx 8$	$F = 19.9$ $F \approx 20$
$v_2 = 5$	$F = 6.6$	$F = 16.3$

It is at once apparent; the interaction A $\times$ B is significant at the 1% level. The main effects A and B can now be compared with the A $\times$ B interaction, but their variance ratios, with only one degree of freedom in numerator and in denominator, come nowhere near the five-percentile value $F = 161.4$ and are several orders of magnitude less than the one-percentile value $F = 4052$:

$$\text{For factor A}, F = \frac{20910.25}{276.125} = 75.72 \text{ with } v_1 = 1 \text{ and } v_2 = 1$$

$$\text{For factor B}, F = \frac{1891.25}{276.125} = 6.84 \text{ with } v_1 = 1 \text{ and } v_2 = 1$$

The result of the experiment can be interpreted as an interaction between factor A and factor B rather than main effects of either factor. This can be represented in a tableau displaying the differences between the average responses at all four treatment combinations of A with B.

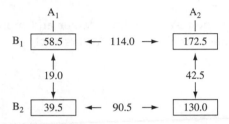

It is unlikely that the interaction effect would have been identified in single-factor experiments unless prior technical knowledge hinted at its existence. Even then, the two-factor experiment with replication was more efficient and effective than a group of separate single-factor experiments.

Clearly, the experiment in eight trials with replication of four treatment combinations was vastly superior in efficiency and effectiveness to the smaller four trial experiment which failed to establish the influence of factor B and only marginally established the significance of factor A. The improvement was due to replication which made it possible to detect the interaction as well as improving sensitivity by avoiding the top line of the F-distribution table when testing for significance.

As an alternative the experiment could have been enlarged by increasing the number of levels of each factor. Three levels in each of two factors without replication would have required nine trials with a total of eight degrees of freedom. Each factor would have had two degrees of freedom, leaving four for the residual. It would not have been possible to examine an interaction. If an interaction had been present it would simply have inflated the residual and reduced the sensitivity of testing the main effects. Replication in eight trials of two factors each at two levels was a better option, though it must be said that with three levels of each factor it would have been possible to explore the possibility of nonlinear effects.

Table 5.7   Four factors without replication

		A_1		A_2	
		B_1	B_2	B_1	B_2
C_1	D_1	38	55	38	53
	D_2	40	46	44	57
C_2	D_1	41	52	38	55
	D_2	59	63	62	64

Table 5.7 records the responses for a larger and more complex experiment involving sixteen trials of four factors each at two levels without replication.

Group totals

A_1 : 394 A_2 : 411 B_1 : 360 B_2 : 445 C_1 : 371 C_2 : 434 D_1 : 370 D_2 : 435

A_1B_1 : 178 A_1B_2 : 216 A_2B_1 : 182 A_2B_2 : 229
A_1C_1 : 179 A_1C_2 : 215 A_2C_1 : 192 A_2C_2 : 219
A_1D_1 : 186 A_1D_2 : 208 A_2D_1 : 184 A_2D_2 : 227
B_1C_1 : 160 B_1C_2 : 200 B_2C_1 : 211 B_2C_2 : 234
B_1D_1 : 155 B_1D_2 : 205 B_2D_1 : 215 B_2D_2 : 230
C_1D_1 : 184 C_1D_2 : 187 C_2D_1 : 186 C_2D_2 : 248

$A_1B_1C_1$: 78 $A_1B_1C_2$: 100 $A_1B_2C_1$: 101 $A_1B_2C_2$: 115
$A_2B_1C_1$: 82 $A_2B_1C_2$: 100 $A_2B_2C_1$: 110 $A_2B_2C_2$: 119
$A_1B_1D_1$: 79 $A_1B_1D_2$: 99 $A_1B_2D_1$: 107 $A_1B_2D_2$: 109
$A_2B_1D_1$: 76 $A_2B_1D_2$: 106 $A_2B_2D_1$: 108 $A_2B_2D_2$: 121
$A_1C_1D_1$: 93 $A_1C_1D_2$: 86 $A_1C_2D_1$: 93 $A_1C_2D_2$: 122
$A_2C_1D_1$: 91 $A_2C_1D_2$: 101 $A_2C_2D_1$: 93 $A_2C_2D_2$: 126
$B_1C_1D_1$: 76 $B_1C_1D_2$: 84 $B_1C_2D_1$: 79 $B_1C_2D_2$: 121
$B_2C_1D_1$: 108 $B_2C_1D_2$: 103 $B_2C_2D_1$: 107 $B_2C_2D_2$: 127

Individuals: 805

Crude sums of squares

$$\text{A } (394^2 + 411^2)/8 = 324157/8 = 40519.6250$$
$$\text{B } (360^2 + 445^2)/8 = 327625/8 = 40953.1250$$
$$\text{C } (371^2 + 434^2)/8 = 325997/8 = 40749.6250$$
$$\text{D } (370^2 + 435^2)/8 = 326125/8 = 40765.6250$$
$$\text{A} \times \text{B}: (178^2 + 216^2 + 182^2 + 229^2)/4 = 163905/4 = 40976.2500$$
$$\text{A} \times \text{C}: (179^2 + 215^2 + 192^2 + 219^2)/4 = 163091/4 = 40772.7500$$
$$\text{A} \times \text{D}: (186^2 + 208^2 + 184^2 + 227^2)/4 = 163245/4 = 40811.2500$$
$$\text{B} \times \text{C}: (160^2 + 200^2 + 211^2 + 234^2)/4 = 164877/4 = 41219.2500$$
$$\text{B} \times \text{D}: (155^2 + 205^2 + 215^2 + 230^2)/4 = 165175/4 = 41293.7500$$
$$\text{C} \times \text{D}: (184^2 + 187^2 + 186^2 + 248^2)/4 = 164925/4 = 41231.2500$$
$$\text{A} \times \text{B} \times \text{C}: (78^2 + 100^2 + 101^2 + 115^2 + 82^2 + 100^2 + 110^2 + 119^2)/2$$
$$= 82495/2 = 41247.5000$$
$$\text{A} \times \text{B} \times \text{D}: (79^2 + 99^2 + 107^2 + 109^2 + 76^2 + 106^2 + 108^2 + 121^2)/2$$
$$= 82689/2 = 41344.5000$$
$$\text{A} \times \text{C} \times \text{D}: (93^2 + 86^2 + 93^2 + 122^2 + 91^2 + 101^2 + 93^2 + 126^2)/2$$
$$= 82585/2 = 41292.5000$$
$$\text{B} \times \text{C} \times \text{D}: (76^2 + 84^2 + 79^2 + 121^2 + 108^2 + 103^2 + 107^2 + 127^2)/2$$
$$= 83565/2 = 41782.5000$$
$$\text{Individuals: } 38^2 + 55^2 + 38^2 + 53^2 + 40^2 + 46^2 + 44^2 + 57^2 + 41^2 + 52^2$$
$$+ 38^2 + 55^2 + 59^2 + 63^2 + 62^2 + 64^2 = 41867.0000$$
$$\text{Correction factor } 805^2/16 = 40501.5625$$

Adjusted sums of squares

$$\text{A } 40519.6250 - 40501.5625 = 18.0625$$
$$\text{B } 40953.1250 - 40501.5625 = 451.5625$$
$$\text{C } 40749.6250 - 40501.5625 = 248.0625$$
$$\text{D } 40765.6250 - 40501.5625 = 264.0625$$
$$\text{A} \times \text{B } 40976.2500 - 40501.5625 - (18.0625 + 451.5625) = 5.0625$$
$$\text{A} \times \text{C } 40772.7500 - 40501.5625 - (18.0625 + 248.0625) = 5.0625$$
$$\text{A} \times \text{D } 40811.2500 - 40501.5625 - (18.0625 + 264.0625) = 27.5625$$
$$\text{B} \times \text{C } 41219.2500 - 40501.5625 - (451.5625 + 248.0625) = 18.0625$$
$$\text{B} \times \text{D } 41293.7500 - 40501.5625 - (451.5625 + 264.0625) = 76.5625$$
$$\text{C} \times \text{D } 41231.2500 - 40501.5625 - (248.0625 + 264.0625) = 217.5625$$
$$\text{A} \times \text{B} \times \text{C}: 41247.5000 - 40501.5625 - (18.0625 + 451.5625 + 248.0625)$$
$$-(5.0625 + 5.0625 + 18.0625) = 0.0625$$
$$\text{A} \times \text{B} \times \text{D}: 41344.5000 - 40501.5625 - (18.0625 + 451.5625 + 264.0625)$$
$$-(5.0625 + 27.5625 + 76.5625) = 0.0625$$

A × C × D: $41292.5000 - 40501.5625 - (18.0625 + 248.0625 + 264.0625)$
$\qquad - (5.0625 + 27.5625 + 217.5625) = 10.5625$
B × C × D: $41782.5000 - 40501.5625 - (451.5625 + 248.0625 + 264.0625)$
$\qquad - (18.0625 + 76.5625 + 217.5625) = 5.0625$
Individuals $41867.0000 - 40501.5625 = 1365.4375$

Note that when calculating the adjusted sums of squares for three-factor interactions the contributions of the relevant two-factor interactions as well as principal factors have to be taken into account. With sixteen trials the sum of squares may be partitioned between six two-factor and three three-factor interactions as well as all four main effects. The complete set is shown in Table 5.8.

The first point to note is that the mean squares for A, A × B, A × C, B × C, A × B × C, A × B × D, A × C × D and B × C × D are less than or equal to the residual mean square. With only one degree of freedom in each it would be difficult to argue that the differences are significant. The sums of squares and the degrees of freedom can be merged with the residual to form a new residual with a sum of squares $(3 \times 18.0625) + (3 \times 5.0625) + (2 \times 0.0625) + 10.5625 = 80.0625$, nine degrees of freedom, and a mean square of 8.896.

Relative to the new mean square the variance ratio for A × D, $F = 27.5625 / 8.896 = 3.10$ is smaller than anything in the first column of the table of percentage points of the F-distribution in Appendix E so that, too, can be

Table 5.8 Preliminary analysis of variance

Source of variation	Sum of squares	Degrees of freedom	Mean square
A	18.0625	1	18.0625
B	451.5625	1	451.5625
C	248.0625	1	248.0625
D	264.0625	1	264.0625
A × B	5.0625	1	5.0625
A × C	5.0625	1	5.0625
A × D	27.5625	1	27.5625
B × C	18.0625	1	18.0625
B × D	76.5625	1	76.5625
C × D	217.5625	1	217.5625
A × B × C	0.0625	1	0.0625
A × B × D	0.0625	1	0.0625
A × C × D	10.5625	1	10.5625
B × C × D	5.0625	1	5.0625
Residual	18.0625	1	18.0625
Total	1365.4375	15	

Table 5.9 Final analysis of variance

Source of variation	Sum of squares	Degrees of freedom	Mean square
B	452	1	452
C	248	1	248
D	264	1	264
B × D	77	1	77
C × D	217	1	217
Residual	107	10	11
Total	1365	15	

merged with the residual to make a final residual mean square of 10.7624 with ten degrees of freedom.

The new format for the analysis of variance with the sums of squares and the mean squares rounded off in a sensible fashion is given in Table 5.9. Note that rounding off at an earlier stage would have been dangerous when calculating small differences between large crude sums of squares and large correction factors.

The variance ratio for the B × D interaction is $F = 77/11 = 7.0$. This value is displayed in a schematic diagram for comparison with values extracted from the table of percentage points of the F-distribution in Appendix E. An estimate of the probability is also given.

$v_1 =$	1	1	1
P (%)	5	$P \approx 3$	1
$v_2 = 10$	$F = 5.0$	$F = 7.0$	$F = 10.0$

The B × D interaction is therefore significant between the 5% and 1% levels. With much higher variance ratios (19.7 to 41.1) all the other sources will be highly significant.

Tableaux of mean responses at each treatment combination demonstrate the interactions. The effects of factors B, and especially C, depend on the level of D (and vice versa).

Critical comments

1. Efficient. The response of every single trial contributed to the evaluation of the effect of each factor and each interaction, whether significant or not.
2. Effective. The clearest possible disentanglement of real effects from random variation using statistical and technical evidence. Detection of nonsignificant effects was just as important as detection of significant effects.
3. Accuracy. Setting out the arithmetical steps in fine detail demonstrated that rounding errors must be avoided because of relatively small differences between large quantities, especially in the smaller sums of squares.
4. The very steep increase in responses towards B_2D_2 and C_2D_2 may be indicative of nonlinear relationships. This could be the subject of further experiments with B, C and D each at three levels (assuming these are quantitative rather than qualitative factors).

By now it will be apparent that factorial experiments tend to be quite large operations, even when each factor is restricted to only two levels. The numbers of trials required for 2, 3, 4, 5... factors in complete factorial experiments are $2^2 = 4, 2^3 = 8, 2^4 = 16, 2^5 = 32$... This can be accommodated in agricultural experiments by dividing a field into more smaller strips, but there may be practical difficulties in an engineering environment in managing the disposition of large numbers of trials in time and space. Resort can be made to *fractional factorial* designs based on ½, ¼, ⅛... of a complete factorial design.

Table 5.10 shows how the responses of the previous experiment might have appeared if a ½ × 2^4 design had been adopted.

Table 5.10 Responses in a fractional factorial design

		A_1		A_2	
		B_1	B_2	B_1	B_2
C_1	D_1	38			53
	D_2		50	41	
C_2	D_1		52	42	
	D_2	58			66

Using the same procedure as before the following sums of squares can be calculated

$$
\begin{array}{ll}
& A \times B : 112.5 \\
A{:}2.0 & A \times C : 8.0 \\
B{:}220.5 & A \times D : 4.5 \\
C{:}162.0 & B \times C : 4.5 \\
D{:}112.5 & B \times D : 8.0 \\
& C \times D : 112.5
\end{array}
$$

Three-factor interactions are not accessible because only one response is available at each treatment combination. The sum of squares for a three-factor interaction would be exactly the same as the total sum of squares for individuals.

However, there is a problem. Not all of the ten sums of squares can be accommodated in the analysis of variance table. With eight responses there are only seven degrees of freedom and so it is not possible to make more than seven independent comparisons. One of these must be reserved for the residual. Which six out of the ten sums of squares should we include?

The best that can be done is shown in Table 5.11, in which the minor sums of squares for A, A × C, A × D, B × C, B × D have been omitted. An interaction effect has been included in recognition of the contribution made by the exceptionally high responses 58 and 66 on the bottom line of Table 5.10. This could be interpreted as either an A × B or C × D interaction which would require further investigation. The final analysis of the fractional factorial experiment is shown in Table 5.11.

The variance ratio for factor D is $F = 112.50/4.83 = 23.3$ with one and three degrees of freedom, the same as for the interaction. This is displayed in a schematic diagram for comparison with values extracted from the table of

Table 5.11   Final analysis of variance

Source of variation	Sum of squares	Degrees of freedom	Mean square
B	220.50	11	220.00
C	162.00	1	162.00
D	112.50	1	112.50
Interaction	112.50	1	112.50
Residual	14.50	3	4.83
Total	622.00	7	

percentage points of the F-distribution in Appendix E and with the estimate $F \approx 15.0$ for $v_1 = 1$, $v_2 = 3$.

v_1	1	1
$P\%$	5	1
$v_2 = 2$	$F = 18.5$	$F = 98.5$
$v_2 = 3$	$F \approx 15.0$ | $F = 23.3$	
$v_2 = 5$	$F = 6.6$	$F = 16.3$

The value of the estimate $F \approx 15.0$ at $v_2 = 3$ has to be rather less than $F = 18.5$ at $v_2 = 2$ and substantially greater than $F = 6.6$ at $v_2 = 5$. The exact value does not matter. The important result is that the value $F = 23.3$ in the analysis of variance is such as to leave little doubt about its significance.

In engineering terms the conclusion from the half-fraction factorial experiment would be the same as for the full factorial experiment in respect of the main effects. Factors B, C and D would all be identified as significant beyond all reasonable doubt. The situation with regard to interaction effects would be less clear. The presence of an interaction would be recognised, but its identity as either A × B or C × D would not be clear on statistical grounds alone. Technical knowledge about the nature of the factors would help. The B × D interaction which was evident in the full factorial experiment would escape notice.

Fractional factorial designs are attractive because they enable many factors to be accommodated in an experiment of reasonable size when a full factorial design would be too large to be practicable. For example, a full factorial experiment for six factors each at two levels would require $2^6 = 64$ trials to accommodate all possible treatment combinations. A quarter fractional factorial experiment would require only 16 trials, but the treatment combinations would have to be chosen in a particular way and each significant effect might have as many as four separate interpretations. This degree of confusion (or *confounding*, to use a statistical term) is often resolved by making a blanket assumption that main single-factor effects are the correct interpretation.

Engineers who balk at the sums of squares number-crunching that is inherent in the analysis of variance of large factorial experiments may find the alternative graphical method of *normal plots* (or the variant *half normal plots*) more attractive. They must, however, come to terms with a different sort of statistical analysis. The method will be demonstrated in Table 5.12 (page 92) by applying it to the responses in Table 5.7 (page 84).

The following symbols are used for the responses at the various treatment combinations:

$$A_1B_1C_1D_1:1 \quad A_2B_2C_1D_1:ab \quad A_2B_2C_2D_1:abc$$
$$A_2B_1C_1D_1:a \quad A_2B_1C_2D_1:ac \quad A_2B_2C_1D_2:abd$$
$$A_1B_2C_1D_1:b \quad A_2B_1C_1D_2:ad \quad A_2B_1C_2D_2:acd$$
$$A_1B_1C_2D_1:c \quad A_1B_2C_2D_1:bc \quad A_1B_2C_2D_2:bcd$$
$$A_1B_1C_1D_2:d \quad A_1B_2C_1D_2:bd \quad A_2B_2C_2D_2:abcd$$
$$A_1B_1C_2D_2:cd$$

Using this notation it can be shown that

$$\text{Effect of A} = \tfrac{1}{8}\,(a-1)(b+1)(c+1)(d+1)$$
$$\text{Effect of B} = \tfrac{1}{8}\,(a+1)(b-1)(c+1)(d+1)$$
$$\text{Effect of C}\times\text{D} = \tfrac{1}{8}\,(a+1)(b+1)(c-1)(d-1)$$
etc.

Expanding these expressions

For A: $-1 + a - b - c - d + ab + ac + ad - bc - bd - cd + abc + abd$
$+acd - bcd + abcd$

For B: $-1 - a + b - c - d + ab - ac - ad + bc + bd - cd + abc + abd$
$-acd + bcd + abcd$

For C $\times$ D: $1 + a + b - c - d + ab - ac - ad - bc - bd + cd - abc - abd$
$+acd + bcd + abcd$

etc.

All of the expressions have the same terms, but different positive and negative signs, All fifteen are embodied in the vertical columns of Table 5.12. The contrast values were calculated using this notation and are given at the bottom of each column.

The underlying principle of normal plotting of factorial experiment responses is that if there are no significant effects the contrast values will behave like sample data from a normal distribution, with mean zero and standard deviation one, and would follow a linear plot on probability graph paper. Departure from linearity is indicative of a significant effect.

To proceed with the analysis the contrast values are ranked in order of magnitude in preparation for plotting against a probability scale. To facilitate this operation Grove and Davis (1992) have provided tables of full normal scores which are, in effect, the expected values of individuals in a sample from a normal distribution. These are displayed on page 93.

Table 5.12 Contrasts for the data in Table 5.7

A	B	C	D	A×B	A×C	A×D	B×C	B×D	C×D	A×B×C	A×B×D	A×C×D	B×C×D	Residual	Responses
−	−	−	−	+	+	+	+	+	+	−	−	−	−	+	..1:38
+	−	−	−	−	−	−	+	+	+	+	+	+	−	−	..a:38
−	+	−	−	−	+	+	−	−	+	+	+	−	+	−	..b:55
−	−	+	−	+	−	+	−	+	−	+	−	+	+	−	..c:41
−	−	−	+	+	+	−	+	−	−	−	+	+	+	−	..d:40
+	+	−	−	+	−	−	−	−	+	−	−	+	+	+	.ab:53
+	−	+	−	−	+	−	−	+	−	−	+	−	+	+	.ac:38
+	−	−	+	−	−	+	+	−	−	+	−	−	+	+	.ad:44
−	+	+	−	−	−	+	+	−	−	−	+	+	−	+	.bc:52
−	+	−	+	−	+	−	−	+	−	+	−	+	−	+	.bd:46
−	−	+	+	+	−	−	−	−	+	+	+	−	−	+	.cd:59
+	+	+	−	+	+	−	+	−	−	+	−	−	−	−	abc:55
+	+	−	+	+	−	+	−	+	−	−	+	−	−	−	abd:57
+	−	+	+	−	+	+	−	−	+	−	−	+	−	−	acd:62
−	+	+	+	−	−	−	+	+	+	−	−	−	+	−	bcd:63
+	+	+	+	+	+	+	+	+	+	+	+	+	+	+	abcd:64
Divisor 8	8	8	8	8	8	8	8	8	8	8	8	8	8	8	
Contrast Values 2.13	10.63	7.88	8.13	1.13	−1.13	2.63	−2.13	−4.38	7.38	−0.13	0.13	−1.63	1.13	−2.13	

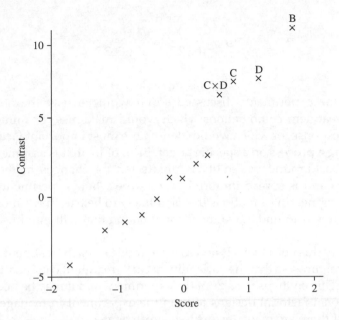

Figure 5.1 Normal plot

Contrast	−4.38	−2.13	−1.63	−1.13	0.13	1.13	1.13	2.13	2.63	7.38	7.88	8.13	10.63
Score		−1.68	−1.16	−0.85	−0.60	−0.39	−0.19	0.00	0.19	0.39	0.60	0.85	1.18	1.68

The normal plot, Figure 5.1, clearly identifies factors B, C, D and the interaction $C \times D$ as being 'out of line' with the others, though the borderline interaction $B \times D$ does not distinguish itself.

It may seem strange that two apparently different methods of analysis should reach the same conclusion, yet both are linked mathematically. Given a set of $(n-1)$ orthogonal contrasts calculated from n responses in a two-level experiment each contrast makes a contribution given by

$$\text{Contribution to sum of squares} = \frac{n}{4} \times (\text{value of contrast})^2$$

(Grove and Davis, 1992). Thus, for A

$$\frac{16}{4} \times 2.125^2 = 18.0625$$

for B

$$\frac{16}{4} \times 10.625^2 = 451.5625$$

Etc. (as in Table 5.8)

All of the experiments discussed so far would satisfy the objective of finding treatment combinations which would maximise (or minimise) the average response, or which would identify the most important factors to use for holding a process on a specific target. In all of them it is assumed that the residual background variability is unaffected by changes in the level of the factors and is spread uniformly across the whole experimental region. In many engineering situations this is unlikely to be true and a more important objective is to find a treatment combination that will minimise random variability.

To satisfy this objective it is necessary to replicate each treatment combination several times so that the variability at each treatment combination can be assessed. This enlarges the size of the experiment and it may be necessary to use fractional factorial designs to keep the experiment in manageable proportions if there are many factors to be investigated.

Taguchi (1985), an engineer, addresses the problem by introducing the 'signal to noise ratio' concept from electronic engineering. His designs also include the novel concept of using two overlapping orthogonal arrays to determine the treatment combinations, one dealing with controllable factors, the other with uncontrollable factors. He also makes use of graphical displays of significant effects. Examples of the application of Taguchi methods are to be found in Quinlan (1985) and Antony & Kaye (1996).

Quinlan describes an investigation of 15 process factors, each of which was tested at two levels. A complete factorial experiment would have required $2^{15} = 32,768$ trials, so a highly fractionated design was adopted using sixteen trials each of which was replicated four times making a total of 64 trials. His comment that 'the conduct of this experiment was not easy' must carry weight.

Engineers embarking on experimental investigations of industrial processes must be prepared to face considerable difficulty in managing and controlling large-scale trials. That is not a criticism of the statistical design of experiments. It is simply recognition of one of the hard facts of life. Before closing this section it must be said there are other ways of investigating complex multivariate processes in manufacturing industry. In a given situation it is the responsibility of the engineer to decide which technique is most suitable – the design of experiments, or something else. Two other techniques which might be considered are *evolutionary operation and multiple regression*.

5.2 Evolutionary Operation

Consider the situation in a glass works accommodating two identical tube drawing machines operating on markedly different production schedules. One machine catering for the fluorescent lamp industry, producing the same size of tubing, night and day, week in week out. The other machine catering for individual customers requiring a wide range of different sizes in terms of diameter and wall thickness necessitating frequent changes of machine settings during a normal working week.

In both cases the size of tubing would be determined by a complex of factors – the temperature in the melting tank, the forehearth temperature, the setting of the feed orifice, the rate of feeding glass onto the hollow mandrel, the speed of rotation of the mandrel and the angle of its setting, the ambient temperature in the vicinity of the mandrel, the pressure of air passing through it into the tubing, the speed of the tractor drawing tubing off the mandrel, etc. All of these factors could be recorded at intervals during production along with precise measurements of dimensions (diameter, ovality, wall thickness) of samples of tubing extracted from the production line. What to do with the data thus collected?

The relationship between the process factors and their effect on size is a complex one. The factors tend to operate against each other and so there is no unique combination for a specific dimension of tubing. The problem would be to find an optimum which would combine maximum production efficiency with minimum variability of dimension.

The fluorescent tubing production line would be ideally suited for evolutionary operation (EVOP) which was developed at Imperial Chemical Industries in the 1950s (Box 1957) and which became the subject of a textbook in its own right (Box and Draper 1969). The method used simple statistical ideas and was based on the philosophy that a process should generate information on how to improve the product as well as generating manufactured product.

In evolutionary operation a carefully planned cycle of minor variants on the normal process is agreed. The production routine then consists of running each of the variants in turn and continually repeating the cycle while data is collected. In the fullness of time the accumulated data will show in which direction to move to secure an improvement. After making a small change in that direction a new cycle of variants is introduced and the process of evolutionary operation continues.

The second production line catering for individual customers requiring many different sizes of tubing in orders of varying amount would cover a wide selection of operating conditions. The data which it generated would be suitable for analysis by the application of multiple regression which is the technique to be discussed in Section 5.3.

In his 1957 article Box made an interesting comparison with the evolution of species. Living things advance by means of two mechanisms:

 (i) genetic variability due to agencies such as mutation;
(ii) natural selection.

He argued that industrial processes advance in a similar manner. The discovery of a new process of manufacture corresponds to a mutation. Adjustment of process factors to their best levels involves natural selection.

Further developments in evolutionary operation continue to be made. A recent publication (Luangpaiboon *et al.* 2000) compares EVOP strategies which include methods with a stochastic element such as simulated annealing (Laarhoven and Aarts 1987, Press *et al.* 1992) with those based on genetic algorithms (Holland 1975).

5.3 Multiple Regression

The underlying principles of multiple regression will be demonstrated using the data in Table 5.13. This data is artificial, having been devised specially for the purpose, but it is realistic. Readers familiar with manufacturing processes will recognise that the dependent variable Y could be a property of a finished product, such as its tensile strength. X might be the proportion

Table 5.13 Process records

Sequence	Y	X	Z
1	151	0.86	957
2	143	0.52	995
3	138	0.48	990
4	136	0.53	956
5	140	0.78	935
6	157	0.96	966
7	133	0.42	973
8	136	0.39	1008
9	128	0.49	945
10	151	0.75	989

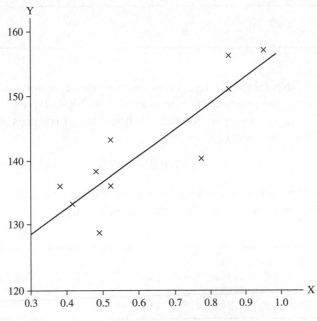

Figure 5.2   Linear regression

of a particular element in the mixture of raw materials. Z could be a process temperature. In statistical jargon Y is the *dependent variable*, X and Z are the *independent variables*.

The situation that is represented in this data is one in which the level of factor X has been used from time immemorial to determine the value of Y to meet customers' requirements. Business is being lost to a competitor who is able to provide better control of Y at a lower cost. What is to be done?

The regression of Y on X carried out as described in Chapter 3, Section 3.6 produces the result depicted in Figure 5.2. Clearly, there is a linear relationship between X and Y, but there is also an uncomfortably large spread of points about the line. The regression equation is

$$Y = 117.3 + 38.8X$$

The inherent *residual error* in using X as a predictor of Y is measured by the vertical intercepts between individual points and the regression line. The standard deviation of this error is $\sigma = 4.52$ and the individual intercepts are as follows;

Sequence	1	2	3	4	5	6	7	8	9	10
Residual	0.3	5.5	2.1	−1.9	−7.6	2.4	−0.6	3.5	−8.3	4.6

It is thought that factor Z also exerts an influence on property Y. It would make good sense to see whether there is a relationship between factor Z and the residuals Y_{res} by using regression analysis. This produces an equation which identifies the effect of Z on Y:

$$Y_{res} = -170.2 + 0.175Z$$

This equation can be used to adjust the recorded values of Y to take account of the contribution of Z by subtracting Y_{res} from Y.

Sequence	1	2	3	4	5	6	7	8	9	10
Y	151.0	143.0	138.0	136.0	140.0	157.0	133.0	136.0	128.0	151.0
Correction	2.5	−4.1	−3.3	2.7	6.4	1.0	−0.3	−6.4	4.6	−3.1
Y_{adj}	153.5	138.9	134.7	138.7	146.4	158.0	132.7	129.6	132.6	147.9

The regression of Y_{adj} on X now yields the effect of X on Y unimpeded by the presence of Z:

$$Y_{adj} = 112.0 + 47.4X$$

The independent effects of X and Z on Y are now known and can be embodied in a single equation:

$$Y = a + 47.4X + 0.175Z$$

where a is the *intercept term* necessary to make the regression pass through the grand average of the original data

$$\overline{Y} = 141.3 \quad \overline{X} = 0.618 \quad \overline{Z} = 971.4$$

This turns out to be $a = -58.1$ and so the final result is:

$$Y = -58.1 + 47.4X + 0.175Z$$

The residuals for the new equation have a standard deviation $\sigma = 1.54$ (compared with $\sigma = 4.52$ previously) and they show a vast overall improvement in prediction errors:

Sequence	1	2	3	4	5	6	7	8	9	10
Residual	0.8	2.3	0.0	1.6	-2.6	0.4	0.9	-0.8	-2.6	0.4

Using X and Y as predictors in planning future production schedules and using a graphical reference chart (Figure 5.3) the hypothetical manufacturer would now have a better chance of keeping his customers happy with closer control of whatever target values of property Y they wish to specify.

He would also be able to reduce his operating costs by not having to scrap or rework batches of product that missed the target, as happened so often in the past. He might even do better than his competitors and win back lost customers!

Fortunately, the toing and froing that has been used in this demonstration is no longer necessary. There is a mathematical solution using matrix algebra that can be applied to any number of independent variables and this is readily available in statistical computer software. This brings with it, however, the danger of producing horrendously incorrect regression equations if the regression technique is not properly understood by the user.

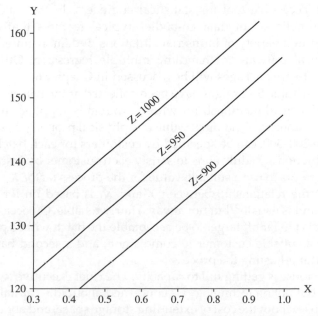

Figure 5.3 Production planning chart

Be warned! Engineers should never treat regression analysis as a sort of mincing machine into which they shovel every scrap of data they can lay hands on in the mistaken belief that whatever is printed out will be utterly reliable. If the independent variables are entered in the wrong order the final equation may be very far from a true representation of the process. If any of the so-called independent variables are highly correlated with one another it is possible for the computer to churn out regression coefficients whose numerical values are so wide of the mark they may even change their sign from positive to negative or vice versa! This is particularly the case if the correlated variables are process factors that have been used to compensate for one another in an endeavour to find a suitable combination for better operating conditions.

In the early days of regression programmes for mainframe computers Hamaker (1962) observed that 'forward selection' or 'backward elimination' of the independent variables gave differing results and opined that 'in industrial applications at least, the question of how many independent variables should be retained is a technical rather than a statistical problem'. That opinion still holds good today, yet there are statistical considerations that should not be ignored.

With the advent of personal computers multiple regression has become a much more accessible tool for statistical engineers. In the following case study a synthetic set of data embodies typical features that have been encountered in a variety of industrial situations. It demonstrates the use of SPSS statistical software for handling multiple regression. Other suitable statistical software packages will be discussed in Chapter 6.

The data in Table 5.14 might have been collected at the behest of the chief executive of a small manufacturer whose product is supplied in batches to individual customers. A nominal value of a physical property Y, somewhere in the range 200–300, can be specified by customers for each batch.

The production schedule aims to satisfy each customer by instructing the process operators to use a suitable value for the process factor X_1. An understanding of the relationship between Y and X_1 is based on long practical experience and is usually (but not always) fairly reliable. On occasions when the product is too far off target to be acceptable the batch will be put in store, waiting for a suitable customer to come along, and a second batch will be prepared after adjusting the process.

The warehouse is getting full to capacity. The chief executive (who is good on finance, but not on technology) is concerned about idle capital stacked in the warehouse, about the cost of extending storage space, and about delays to customers who are threatening to look for another supplier. He has discussed

Table 5.14 Process Data

Batch	Y	X_1	X_2	X_3	X_4	X_5	X_6
1	255	14.4	0.50	0.55	0.60	16	18
2	295	19.3	0.52	0.45	0.58	17	14
3	213	10.9	0.47	0.53	0.40	8	4
4	213	12.3	0.36	0.62	0.33	9	12
5	246	15.5	0.34	0.70	0.35	12	17
6	236	12.7	0.53	0.50	0.53	14	22
7	257	13.2	0.67	0.38	0.71	18	6
8	223	12.1	0.44	0.52	0.40	14	25
9	271	14.8	0.71	0.27	0.68	12	13
10	269	17.4	0.39	0.68	0.34	15	15
11	277	17.3	0.54	0.42	0.48	6	17
12	212	11.2	0.43	0.56	0.39	6	11
13	247	14.6	0.45	0.57	0.48	16	14
14	275	17.7	0.42	0.66	0.43	18	6
15	286	19.5	0.37	0.64	0.42	16	8
16	243	13.8	0.51	0.45	0.48	12	22
17	258	15.7	0.43	0.63	0.47	10	10
18	271	16.6	0.52	0.53	0.50	21	14
19	213	11.0	0.44	0.55	0.40	20	22
20	282	18.5	0.44	0.60	0.44	13	19

the problem with an old school friend at a conference and has been given the advice, 'Record everything and use multiple regression.' The factors X_2, X_3, X_4, X_5, X_6 identify everything that he can think of in consultation with the production scheduler and the process operators. Records are then kept for a series of 20 batches.

The first task in multiple regression is to ensure that the data has been correctly entered. SPSS software offers a variety of facilities that give the user a broad overview of the data. There is a case processing summary which draws attention to missing entries. There are tables of descriptive statistics for each variable (minimum, maximum, mean, median, standard deviation, etc.) and the user can view on the screen *stem-and-leaf plots* and/or *box plots* for each variable in turn.

A box plot displays the median value, the upper and lower 25 percentiles, and the extremities of a coherent distribution with similar properties to the data. The box plot for X_4 is shown in Figure 5.4. It draws attention to batch 7 in which the value $X_4 = 0.71$ appears to be unusually large. Is it an error, and if not, is it important? In fact, it is not an error and it is important, so it is

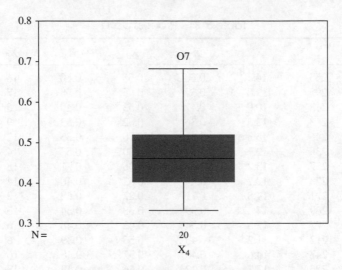

Figure 5.4   Box plot for variable X_4

retained in the analysis. In contrast, if the value $X_5 = 81$ had been entered instead of $X_5 = 18$ it would have stuck out like a sore thumb in the box plot for X_5. There would then have been the opportunity to check the error and correct it, rather than eliminate all the data for batch 7 which would have lost the important value $X_4 = 0.71$.

Stem-and-leaf plots reveal the shape of the distribution of data. The stem-and-leaf plot for X_4 is as follows:

X4 Stem-and-leaf plot	
Frequency	Stem and leaf
2.00	3.34
2.00	3.59
6.00	4.000234
4.00	4.7888
2.00	5.03
1.00	5.8
1.00	6.0
1.00	6.8
1.00	Extremes ($> = 0.71$)
Stem width	0.10
Each leaf	1 case (s)

This histogram is constructed automatically without the user having to make decisions about class intervals and class boundaries. Take note, in the X_4 column of Table 5.14, there are three values at 0.40, and one each at 0.42, 0.43 and 0.44 making a total frequency of 6.00. The 'stem' is 4 and the 'leaves' are 0, 0, 0, 2, 3 and 4, as shown above. It is at once apparent, the value $X_4 = 0.71$ is sitting on the tail of a skewed distribution and is not a rank outsider. Before proceeding with regression analysis it is advisable to examine the patterns of association between the dependent and (so-called) independent variables as well as those that may exist amongst the latter. SPSS software facilitates this in two ways – by displaying (a) a matrix of miniature scatter diagrams (each one slightly larger than a postage stamp); and (b) a table of correlation coefficients for each variable taken in conjunction with every other one in turn.

The scatter diagrams are shown in Figure 5.5. As might be expected the scatter diagrams for Y display a strong association with X_1, but nothing very impressive with the other independent variables. The X_2, X_3, X_4 trio appear to be strongly correlated with one another. The X_5, X_6 pair do not exhibit any correlations.

Reference to the extensive table of correlation coefficients on the screen yields the following selection of items of interest:

Table 5.15 Correlation coefficients

Pair	Coefficient
Y with X_1	0.944
Y with X_2	0.230
Y with X_3	−0.093
Y with X_4	0.395
X_2 with X_3	−0.933
X_3 with X_4	−0.779
X_4 with X_2	0.904

The evidence now suggests three regression models that would be worth considering:

Model 1	Regression of Y on X_1 to evaluate what the production scheduler is currently doing
Model 2	Regression of Y on X_1 and X_4, the latter being the variable in the intercorrelated group most strongly associated with Y
Model 3	Regression of Y on all variables from $X_1 - X_6$

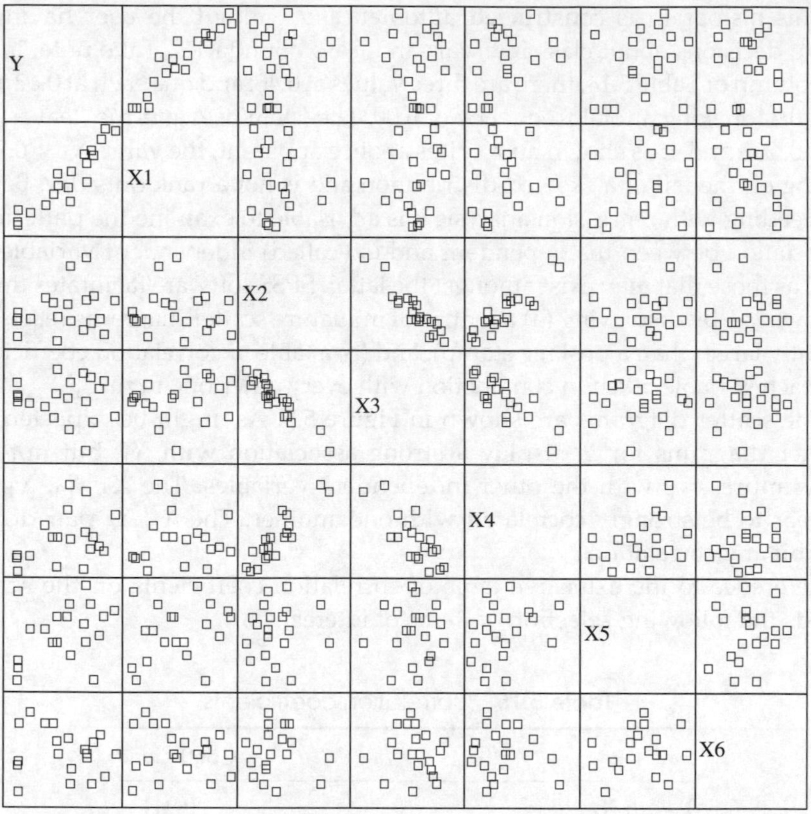

Figure 5.5 Scatter diagrams for all variables

SPSS software obligingly (and instantly) provides regression coefficients along with the standard error of the Y predictor and an analysis of variance for each model.

1. Bivariate regression

$$Y = 116.4 + 9.09X_1$$
Standard error of prediction $= 9.05$

	Sum of squares	Degrees of freedom	Mean square	F
Regression	12067.7	1	12067.7	147
Residual	1474.0	18	81.9	
Total	13541.8	19		

2. Trivariate regression

$$Y = 82.9 + 8.83X_1 + 79.6X_4$$
Standard error of prediction $= 2.83$

	Sum of squares	Degrees of freedom	Mean square	F
Regression	13405.6	2	6702.82	837
Residual	136.2	17	8.01	
Total	13541.8	19		

3. Multivariate regression

$$Y = 34.6 + 9.09X_1 + 22.9X_2 + 106X_3 + 36.5X_4 + 0.0321X_5 + 0.0516X_6$$
Standard error of prediction $= 1.44$

	Sum of squares	Degrees of freedom	Mean square	F
Regression	13514.9	6	2252.48	1088
Residual	26.9	13	2.07	
Total	13541.8	19		

The model which is finally selected is the trivariate version enabling Y to be predicted on the basis of values of X_1 and X_4. This gives a dramatic improvement of prediction with the standard error reduced from 9.05 to 2.83, and it only requires one more variable to be taken into account when scheduling production. There will now be a very good chance of meeting the customer's requirement within reasonable limits on every occasion.

The multivariate model embracing all the recorded variables is rejected for two reasons. First, because the effectiveness of the trivariate model renders the monitoring of additional variables unnecessary. Second, because significance tests displayed on the computer screen show that the coefficients for X_5 and X_6 are not significant. Subroutines in SPSS compute the probability value and display this alongside the numerical value of statistics such as t and F.

The Y/X_1 scatter plot in Figure 5.6 demonstrates the full extent of the problem that the production schedulers were faced with when predicting on the basis of X_1 alone.

Figure 5.7 is a prediction chart which will assist them to incorporate the X_4 variable in their predictions and so greatly enhance the prospect of customer

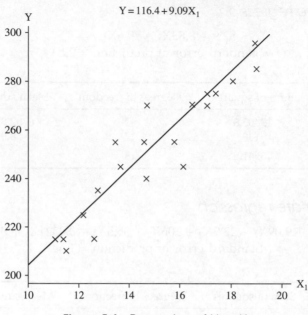

Figure 5.6   Regression of Y on X_1

satisfaction as well as relieving the chief executive of the financial burden of unnecessary storage of production failures. Arrangements will, of course, have to be made for the variable X_4 to be measured and recorded as a matter of routine.

It is good practice to look at the pattern of residuals before logging out of regression analysis, especially if the residual variance is uncomfortably high. This may reveal circumstances which were overlooked at the outset. It might be a mistake to apply linear regression to data embodying nonlinear relationships. The system that generated the data may be influenced by factors that were not taken into account when the original plan for collecting data was formulated. The set of residuals can be subjected not only to visual examination, but any form of statistical analysis that is appropriate.

A residual r is defined as the difference between the recorded value Y and the predicted value $\hat{Y}$ arrived at by entering the corresponding values of the independent variables recorded in the data. The residuals of the application of trivariate regression to the data in Table 5.14 are recorded in Table 5.16 using the formula $\hat{Y} = 82.9 + 8.83X_1 + 79.6X_4$

The first, and most obvious, comment is that the residuals are distributed more or less symmetrically about zero within the range ± 5. This is an

$$Y = 82.9 + 8.83X_1 + 79.6X_4$$

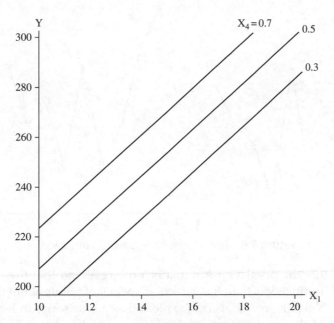

Figure 5.7   Production scheduling chart

Table 5.16   Residuals

Batch	1	2	3	4	5	6	7	8	9	10
X_1	14.4	19.3	10.9	12.3	15.5	12.7	13.2	12.1	14.8	17.4
X_4	0.60	0.58	0.40	0.33	0.35	0.53	0.71	0.40	0.68	0.34
$\hat{Y}$	258	299	211	218	248	237	256	222	268	264
Y	255	295	213	213	246	236	257	223	268	269
R	−3	−4	2	−5	−2	−1	1	1	0	5

Batch	11	12	13	14	15	16	17	18	19	20
X_1	17.3	11.2	14.6	17.7	19.5	13.8	15.7	16.6	11.0	18.5
X_4	0.48	0.39	0.48	0.43	0.42	0.48	0.47	0.50	0.40	0.44
$\hat{Y}$	274	213	250	273	289	243	259	269	212	281
Y	277	212	247	275	286	243	258	271	213	282
r	3	−1	−3	2	−3	0	−1	2	1	1

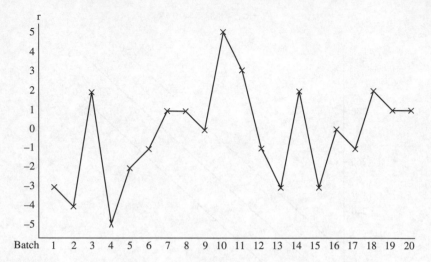

Figure 5.8 Time series of regression residuals

indication that the prediction formula is 'on target' and that the prediction errors will be much less than the vertical dispersal of points about the line on Figure 5.6. Next, the residuals can be plotted as a time series in Figure 5.8.

Because of the gradual ascent of r from batch 4 to batch 10 there is a suspicion that the distribution of r is not random. Could this be associated with the magnitude of Y? Apparently not, judging by Figure 5.9. Perhaps it is the effect of another variable that was not identified and recorded?

The decision whether to pursue this investigation further would have to rest on technical and managerial grounds. Is the improvement in accuracy of production scheduling by introducing X_4 to the prediction formula good enough to maintain competitiveness? Can other suppliers do better? Are customers happy with the new situation? How much will it cost to continue the investigation, etc.?

Before leaving this case study it is worth lingering to consider the interpretation of the regression coefficients. Do they, or do they not, give an accurate representation of the influence of the variables in the manufacturing process? The answer is, yes and (wait for it) no!

In constructing the data set in Table 5.14, the variables X_2, X_3 and X_4 were deliberately correlated with one another, but not with X_1. The variables X_5 and X_6 were purely random and uncorrelated. Values of Y were generated using the formula

$$Y = 40 + 9X_1 + 100X_2 + 30X_3 + 30X_4 + \text{(small element of randomness)}$$

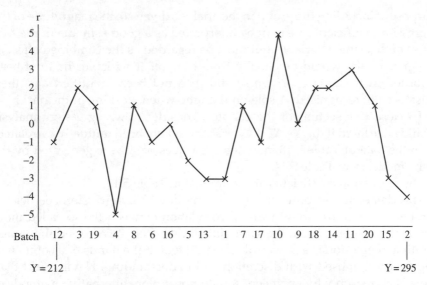

Figure 5.9 Residuals in order of magnitude of Y

Now compare the regression coefficients with the coefficients in the generating formula (Table 5.17):

Table 5.17

	X_1	X_2	X_3	X_4	X_5	X_6
Generator	9.00	100.0	30.0	30.0		
Bivariate	9.09					
Trivariate	8.83			79.6		
Multivariate	9.09	22.9	106.0	36.5	0.0321	0.0516

The coefficients for X_5 and X_6 in the multivariate regression were not significant. In all three regressions the X_1 coefficients can be regarded as good estimators of the coefficient in the generating formula. In the trivariate and multivariate regressions the X_2, X_3 and X_4 coefficients are in wild disagreement with the generating formula. This confirms the danger that was mentioned earlier, of working with so-called independent variables that are highly correlated with one another.

A good working rule is that if a group of 'independent' variables is seen to be highly correlated the one which is most strongly correlated with the dependent variable can be selected as a representative 'leader' for the purpose of including the group in the analysis. Even so, the magnitude of the regression coefficient must not be interpreted as a good estimate of the true effect of that single variable, but must be regarded as the combined effect of the group. That would be equally true if the other variables had not been recorded in the data – or, worse still, had not been identified and their existence not recognised. A solemn thought, when working in the dark!

To close this section of the chapter, consider how regression analysis would have fared if the X_2, X_3, X_4 variables had not been mutually correlated, as in the data of Table 5.18 in which the values of Y were generated by the same model as in Table 5.14.

The regression coefficients are compared in Table 5.19.

This demonstrates quite clearly the way in which regression coefficients can be distorted if there is a strong correlation between the so-called independent variables. Yet, at the same time it can safely be asserted that linear multiple regression is a powerful statistical tool in the hands of a competent engineer if it is used with discretion and understanding. Having said that, there is no reason why engineers should not explore alternative methods of

Table 5.18

Batch	Y	X_1	X_2	X_3	X_4	X_5	X_6
1	224	14.4	0.17	0.51	0.72	16	18
2	255	19.3	0.21	0.52	0.22	17	14
3	242	10.9	0.69	0.54	0.61	8	4
4	256	12.3	0.70	0.58	0.70	9	12
5	253	15.5	0.50	0.58	0.19	12	17
6	199	12.7	0.29	0.50	0.09	14	22
7	240	13.2	0.58	0.33	0.47	18	6
8	274	12.1	0.78	0.59	0.91	14	25
9	223	14.8	0.19	0.50	0.47	12	13
10	260	17.4	0.25	0.58	0.60	15	15
11	252	17.3	0.30	0.37	0.45	6	17
12	242	11.2	0.80	0.38	0.37	6	11
13	268	14.6	0.58	0.31	0.96	16	14
14	284	17.7	0.58	0.67	0.20	18	6
15	320	19.5	0.43	0.53	0.85	16	8
16	250	13.8	0.45	0.50	0.86	12	22
17	309	15.7	0.85	0.70	0.59	10	10
18	233	16.6	0.19	0.47	0.39	21	14
19	192	11.0	0.35	0.42	0.14	20	22
20	315	18.5	0.80	0.32	0.64	13	19

Table 5.19

	X_1	X_2	X_3	X_4	X_5	X_6
Generator	9.00	100.0	30.0	30.0		
Correlated X_2, X_3, X_4	9.09	22.9	106.0	36.5	0.0321	0.0516
Uncorrelated X_2, X_3, X_4	9.57	98.4	36.6	37.2	0.0874	−0.146

Table 5.20

	X_1	X_2	X_3	X_4
Generator	9.00	100.0	30.0	30.0
SPSS	9.09	22.9	106.0	36.5
AMOS	9.08	103.19	34.83	23.94

dealing with multivariate data. It is interesting to record that when the data in Table 5.14, was exposed to AMOS software (Analysis of Moment Structures) it yielded a set of coefficients that were remarkably close to those in the generating model in spite of the correlations in the X_2, X_3, X_4 variables. These are compared in Table 5.20.

Does this not suggest an area of exploration for statistically minded academic engineers?

5.4 More Statistical Methods

The three methods introduced at basic level in this chapter (design of experiments, evolutionary operation, and multiple regression) are not by any means the only statistical methods available to engineers engaged in research and development. All three have been taken further and there are many others that have not yet been mentioned in this text. The problem facing an engineer who is not familiar with applied statistics is knowing where to look, and what to look for. It is not very helpful to point to the titles of over 120 books in Appendix B and a long list of journal titles in Appendix C without offering some guidance.

The guidelines in Appendix A are based on a survey of the papers presented at the *Industrial Statistics in Action 2000 Conference*. This international event at the University of Newcastle upon Tyne was attended by 200 delegates from all over the world. As well as 5 keynote presentations, 60 individual papers were presented, covering virtually every statistical technique currently being applied in industrial situations. Just under 50 distinct areas of application were identified.

All five of the keynote speakers have made distinguished contributions to the development and application of statistical methods in engineering and science. This was reflected in the content of their conference presentations:

Shin Taguchi, President of the American Supplier Institute, USA – *Robust Engineering, The New Engineering Paradigm.*

C. F. Jeff Wu, H C Carver Professor, Department of Statistics and of Industrial and Operations Engineering, University of Michigan, USA – *Experimental Design in the New Millennium.*

Søren Bisgaard, Professor of Quality Management and Technology, University of St. Gallen, Switzerland – *Quality Management and Applied Statistics: Current Developments and Future Trends.*

George Box, Emeritus Professor and Research Director of the Center for Quality and Productivity, University of Wisconsin, USA – *Statistics for Discovery.*

Douglas C. Montgomery, Professor of Engineering, Arizona State University, USA – *Opportunities and Challenges for Industrial Statisticians.*

Most of the other papers published in the Proceedings (Coleman *et al.* 2000) were amply furnished with bibliographic references leading to a wide range of other relevant published sources. The Proceedings can be regarded as an entrée to the whole field of applied statistics. Not all of the papers were necessarily of direct interest to every engineer, but the guidelines in Appendix A will help readers to locate suitable sources without difficulty once they have made up their mind what they are looking for.

Engineers should not be put off by the fact that the reference lists in the Proceedings identify over 120 titles in Appendix B. Many are of a specialist nature and were referred to just once in the context of a paper on a particular subject. Books that were referred to repeatedly by several speakers are the ones most likely to interest engineers because of their general importance.

The following titles identify with statistical methods generally:

6 references: Box, G.E.P., Hunter, W.G. and Hunter, J.S. 1978: *Statistics for Experimenters.* New York: Wiley.
6 references: Montgomery, D.C. 1991: *Introduction to Statistical Quality Control* (2nd edn) New York: Wiley.
5 references: Wetherill, G.B. and Brown, D.W. 1991: *Statistical Process Control.* London: Chapman and Hall.

4 references: Wheeler, D. 1993: Understanding Variation – The Key to Managing Chaos. SPC Pres.

3 references: Bissel, D. 1994: *Statistical Methods for SPC and TQM*. London: Chapman & Hall.

3 references: Box, G.E.P. and Luceño, A. 1997: *Statistical Control by Monitoring and Feedback Adjustment*. New York: Wiley.

3 references: Hines, W.H. and Montgomery, D.C. 1990: *Probability and Statistics in Engineering and Management Science*. New York: Wiley.

3 references: Shewhart, W.A. 1931: *Economic Control of Quality of a Manufactured Product*. New York: Van Nostrand.

3 references: Taguchi, G. 1986: *Introduction to Quality Engineering*. Tokio: Asian Productivity Association.

2 references: Metcalfe, A.V. 1994: *Statistics in Engineering – A Practical Approach*. London: Chapman and Hall.

The following titles identify with design of experiments:

9 references: Grove, D.M. and Davis, T.P. 1992: *Engineering Quality and Experimental Design*. London: Longman.

6 references: Box, G.E.P., Hunter, W.G. and Hunter, J.S. 1978: *Statistics for Experimenters*. New York: Wiley.

3 references: Montgomery, D.C. 2001: *Design and Analysis of Experiments* (5th edn) New York: Wiley.

2 references: Condra, L.W. 1995: *Value Added Management by Design of Experiments*. London, Chapman & Hall.

2 references: Cornell, J. A. 1990: *Experiments with Mixtures*. New York: Wiley.

2 references: Fisher, R.A. 1960: *The Design of Experiments* (7th edn) Edinburgh: Oliver & Boyd.

2 references: Wu, C.J.F. and Hamada, M. 2000: Experiments: Planning, Analysis, and Parameter Design Optimization. New York: Wiley.

Four more titles of interest to statistical engineers:

5 references: Deming, W.E. 1986: *Out of the Crisis*. Cambridge, MA: MIT Press.

4 references: Senge, P.M. 1990: The Fifth Discipline: The Art and Practice of the Learning Organisation. New York: Doubleday.

2 references: Draper, N.R. and Smith, H. 1998: *Applied Regression Analysis*
               (3rd edn) *New York: Wiley*.
2 references: Myers, R.H. and Montgomery, D.C. 1995: Response Surface
               Methodology: Process and Product Optimisation using
               Designed Experiments. New York: Wiley.

Two new titles published since the conference are also important to engineers:

Booker, J.D., Raines, M. and Swift, K.G. 2001: *Designing Capable and Reliable Products*. London: Butterworth-Heinemann.

O'Connor, P.D.T. 2001: Test Engineering - A Concise Guide to Cost-effective Design, Development and Manufacture. Chichester: Wiley.

Engineers pursuing special interests will find plenty to attract their attention, not only in individual conference papers, but also in the extensive lists of bibliographic references. Two examples will be given here to illustrate the point – capital goods and composite materials.

The guideline on capital goods in Appendix A identifies two papers emanating from the Department of Mechanical, Materials and Manufacturing Engineering and the Industrial Statistics Research Unit at the University of Newcastle upon Tyne: *Planning operation start times for the manufacture of capital products with uncertain processing times and resource constraints* by Song, Hicks and Earl, along with *Applying designed experiments to optimise the performance of genetic algorithms used for scheduling complex products in the capital goods industry* by Pongcharoen, Stewardson, Hicks and Braiden. In both papers the theoretical development is supported by case studies conducted with industrial collaborators. Taken together the reference lists run to 47 items, of which only 6 are textbooks. The remainder form a miscellany of publications, not only in statistical journals, but in other professional literature as well.

The guideline on composite materials in Appendix A takes the reader to a paper from Swedish sources entitled *Identification of factors influencing dispersion in split-plot experiments* by Arvidsson, Kammerlind, Hynen and Bergman. The theoretical development of experimental design was supported by a practical case study in which an experiment was conducted with the aim of improving the robustness of the manufacturing process of composite material for military fighter aeroplanes. Again, this was a collaborative effort between academe and industry. The reference list of 20 items was not dominated by statistical textbooks, but identified sources in a variety of professional journals.

Finally, stepping aside from the conference proceedings, the list of journals in Appendix C provides an alternative area of statistical interest for engineers to explore.

Bibliography

Antony, J. and Kaye, M. 1996: Optimisation of core tube life using Taguchi experimental design methodology. *Quality World Technical Supplement*, March, 42–50. London: Institute of Quality Assurance.

Arvidsson, M., Kammerlind, P., Hynen, A. and Bergman, B. 2000: Identification of factors influencing dispersion in split-plot experiments. In Coleman, S., Stewardson, D. and Fairbairn, L. (eds) *Proc. Industrial Statistics in Action 2000 Conference*, vol.I, pp.290–303. University of Newcastle upon Tyne.

Bisgaard, S. 2000: Quality Management and Applied Statistics: Current Developments and Future Trends. In Coleman, S., Stewardson, D. and Fairbairn, L. (eds) *Proc. Industrial Statistics in Action 2000 Conference*, vol.I, pp.12–22. University of Newcastle upon Tyne.

Bissel, D. 1994: *Statistical Methods for SPC and TQM*. London: Chapman & Hall.

Booker, J.D., Raines, M. and Swift, K.G. 2001: *Designing Capable and Reliable Products*. London: Butterworth-Heinemann.

Box, G.E.P. 1957: Evolutionary Operation – A method for increasing industrial productivity. *Applied Statistics* **VI**, 2, 81–101. London: Royal Statistical Society.

Box, G.E.P. and Draper, N.R. 1969: Evolutionary Operation – A Statistical Method for Process Improvement. New York: John Wiley.

Box, G.E.P., Hunter, W.G. and Hunter, J.S. 1978: *Statistical for Experimenters*. New York: Wiley.

Box, G.E.P. and Luceño, A. 1997: *Statistical Control by Monitoring and Feedback Adjustment*. New York: Wiley.

Box, G. 2000: Statistics for Discovery. In Coleman, S., Stewardson, D. and Fairbairn, L. (eds) *Proc. Industrial Statistics in Action 2000 Conference*, vol.I, pp.23–35. University of Newcastle upon Tyne.

Coleman, S., Stewardson, D. and Fairbairn, L. (eds) 2000: *Proc. Industrial Statistics in Action 2000 Conference*, vol.I and II. University of Newcastle upon Tyne.

Condra, L.W. 1995: *Value Added Management by Design of Experiments*. London, Chapman & Hall.

Cornell, J.A. 1990: *Experiments with Mixtures*. New York: Wiley.

Deming, W.E. 1986: *Out of the Crisis*. Cambridge, MA: MIT Press.

Draper, N.R. and Smith, H. 1998: *Applied Regression Analysis* (3rd edn) New York: Wiley.

Fisher, R.A. 1960: *The Design of Experiments* (7th edn) Edinburgh: Oliver & Boyd.

Grove, D.M. and Davis, T.P. 1992: *Engineering Quality and Experimental Design*. London: Longman.

Hamaker, H.C. 1962: On multiple regression analysis. *Statistica Neerlandica* **16**, 31–56.

Hines, W.H. and Montgomery, 1990: *Probability and Statistics in Engineering and Management Science.* New York: Wiley.

Holland, J.H. 1975: *Adaptation in Natural and Artificial System.* Ann Arbor: The University of Michigan Press.

Laarhoven, Van, P.J.M. and Arrts, E.H.L. 1987: *Simulated Annealing, Theory and Applications.* Dordrecht Lancaster Reidel.

Metcalfe, A.V. 1994: *Statistics in Engineering – A Practical Approach.* London: Chapman and Hall.

Luangpaiboon, P., Metcalfe. A.V., Rowlands, R.J., Tham, M.T. and Willis, M.J. Comparison of the Modified Simplex Method and a Genetic Algorithm for Optimising a Chemical Process. In Coleman, S., Stewardson, D. and Fairbairn, L. (eds) *Proc. Industrial Statistics in Action 2000 Conference*, vol.I, pp.36–53. University of Newcastle upon Tyne.

Montgomery, D.C. 2001: *Design and Analysis of Experiments.* (5th edn) New York: John Wiley.

Montgomery, D.C. 1991: *Introduction to Statistical Quality Control* (2nd edn) New York: Wiley.

Montgomery, D.C. 2000: Opportunities and Challenges for Industrial Statisticians. In Coleman, S., Stewardson, D. and Fairbairn, L. (eds) *Proc. Industrial Statistics in Action 2000 Conference*, vol.I, pp.36–53. University of Newcastle upon Tyne.

Myers, R.H. and Montgomery, D.C. 1995: *Response Surface Methodology: Process and Product Optimisation using Designed Experiments.* New York: Wiley.

O'Connor, P.D.T. 2001: *Test Engineering – A Concise Guide to Cost-effective Design, Development and Manufacture.* Chichester: John Wiley.

Pongcharoen, P., Stewardson, D.J., Hicks, C., and Braiden, P.M. 2000: Applying designed experiments to optimise the performance of genetic algorithms used for scheduling complex products in the capital goods industry. In Coleman, S., Stewardson, D. and Fairbairn, L. (eds) *Proc. Industrial Statistics in Action 2000 Conference*, vol.II, pp.98–112. University of Newcastle upon Tyne.

Press, W.H., Vetterling, W.T., Tenkolsky, S.A. and Flannery, B.P. 1992: *Numerical Recipes in Fortran. The Art of Scientific Computing.* Cambridge University Press.

Quinlan, J. 1985: Product improvement by application of Taguchi methods. *Third Supplier Symposium on Taguchi Methods.* Dearborn, MI: American Supplier Institute, Inc.

Senge, P.M. 1990: *The Fifth Discipline: The Art and Practice of the Learning Organisation.* New York: Doubleday.

Shewhart, W.A. 1931: *Economic Control of Quality of a Manufactured Product.* New York: Van Nostrand.

Song, D., Hicks, C. and Earl, C.F. 2000: Planning operation start times for the manufacture of capital products with uncertain processing times and resource constraints. In Coleman. S., Stewardson, D. and Fairbairn, L. (eds) *Proc. Industrial Statistics in Action 2000 Conference*, vol.I, pp.138–152. University of Newcastle upon Tyne.

Taguchi, G. and Wu, Y. 1985: *Introduction to Off-line Quality Control*. Nagoya: Central Japan Quality Control Association.

Taguchi, G. 1986: *Introduction to Quality Engineering*. Tokio: Asian Productivity Association.

Taguchi, S, 2000: Robust Engineering, the New Engineering Paradigm. In Coleman, S., Stewardson, D. and Fairbairn, L. (eds) *Proc. Industrial Statistics in Action 2000 Conference*, vol.I, pp.1–4. University of Newcastle upon Tyne.

Wetherill, G.B. and Brown, D.W. 1991: *Statistical Process Control*. London: Chapman & Hall.

Wheeler, D. 1993: *Understanding Variation – The Key to Managing Chaos*. SPC Press.

Wu, C.F.J. 2000: Experimental Design in the New Millennium. In Coleman, S., Stewardson, D. and Fairbairn, L. (eds) *Proc. Industrial Statistics in Action 2000 Conference*, vol.I, pp.5–11. University of Newcastle upon Tyne.

Wu, C.J.F. and Hamada, M. 2000: *Experiments: Planning, Analysis, and Parameter Design Optimization*. New York: Wiley.

6

Background

Statistical engineering must not be regarded as something so specialised it can stand alone, independent of everything else in the environment in which it operates. On the contrary, it is intimately involved with all that is happening. Three important issues need to be considered: *Measurement* is the origin of the data used in statistical engineering analysis; *Computing*, because of the prevalence of statistical software readily available on PCs; *Management*, because the quality of manufactured products that is served by statistical engineering is dependent on the trinity of management, technology and statistics. Measurement and computing will be discussed in this chapter. Management will be accorded a chapter in its own right.

6.1 Measurement

The data that are used in statistical engineering originate in a process of measurement (using that word in the widest sense to include human observation as well as instrumentation). It should never be taken for granted. There is a particular responsibility on engineers to ensure data is as free as possible from error, especially if instrumentation is the source.

To avoid misunderstanding it should be noted the word *error* tends to have different meanings when used by engineers or by statisticians. To an engineer an error implies that an instrument has not been correctly calibrated. A single reading will be erroneous and a series of readings will

Statistics for Engineers: an Introduction   S.J. Morrison
© 2009 John Wiley & Sons, Ltd

have a mean value that differs from the correct value. In statistics the word 'error' is used in a rather different way to identify the variability or dispersal of individuals. When engineers are in communication with statisticians it is important that the two distinct meanings are not confused.

The sources and types of error, in the engineering sense, are as wide as technology itself. A useful work of reference is Polak and Pande (1999). To quote from the foreword, 'This book sets out to cover most of the types of measurements regularly used by engineers'. The subheading in the title *'Methods and Intrinsic Errors'* conveys the main thrust of the book, which is to alert the engineer to the wide variety of measurement methods and the ways in which they can go wrong. The errors which are covered include not only the inaccuracies of instruments themselves, but the more serious systematic errors intrinsic to the method, which may be very large.

The book amounts almost to an encyclopaedia of methods and errors. It is aimed at a broad spectrum of engineering including manufacturing and process industries as well as research and development. All the examples are real and many of the errors described are large, ranging from 10% to a factor of two or more.

In one of the earlier passages the point is made that nearly all measurements involve humans. Even when their function has been replaced by a computer or robot the human is involved in the selection of the technique and in the interpretation of the result. The human element, ranging from simple mistakes to misinterpretations and confusion caused by badly presented data or instructions, is dealt with in the opening passages. Subsequent chapters deal with specific measurement functions:

Chapter 3. Position, speed and acceleration
Chapter 4. Force, torque, stress and pressure
Chapter 5. Temperature
Chapter 6. Fluid flow
Chapter 7. Electrical measurements and instrumentation
Chapter 8. Properties of materials
Chapter 9. Surface profile, friction and wear
Chapter 10. Internal combustion engine testing
Chapter 11. Computers

The final chapter on computers is of particular interest in the context of statistical engineering. Under the heading of control and data acquisition it deals with the limitations of off-the-shelf packages, with purpose-built computer control systems, with types of control and test equipment, with data acquisition and storage, and with calibration and limitations of a digital system. Under the heading of computer simulation it gives examples of an optical system, of the release of fission gases in nuclear fuel rods, of structural finite element analysis, and of a pneumatic dashpot.

The text ends with a useful bibliography and a brief appendix of definitions of instrument terms.

Useful guidance on measurement can also be found in chapters contributed by Lynn and by Penney to *Research Methods – Guidance for Postgraduates*, (Greenfield 2002). In a chapter entitled *Principles of sampling* Lynn discusses the statistical aspects of sampling. These are particularly relevant in survey data, but they can also be important in technical data if there are reasons (such as cost, meeting a deadline, or the destruction of individuals) for not including every possible unit of measurement.

To quote from the summary, 'Sampling is a complex discipline, yet it is of primary importance in many studies. It is the foundation on which much study is built'. As well as random sampling consideration is given to systematic sampling, stratification, unequal sampling fractions, multistage sampling, capture-recapture sampling, and adaptive sampling. The chapter is supported by a bibliography of 23 items.

In contrast, Penney's chapter *Instrumentation for experiments* deals with electrical measurements. He presents an interesting hierarchic diagram which identifies different forms of digital and analogue instruments for providing characteristic information on the behaviour of technical systems and processes. The terminology of transducers, conditioners, amplifiers, isolators, filters, exciters, etc., is explained. Of particular interest to PC users is the mention of plug-in data acquisition cards (DAQ) and a general purpose interface bus (GPIB) for connecting stand-alone instruments to a PC.

6.2 Statistical Computing

When the Royal Statistical Society was founded in 1834 interest was focused initially on the recording and examination of social statistics. The development of analytical methods which would be applicable in the wider fields of science and technology did not gather momentum until the turn of the century. Even then, powerful methods such as multiple regression depended on the presence of a mathematician. Computation was an obstacle to the application of statistical methods in other professions. Any intelligent person could add a column of figures and come up with the average, but sums of squares were not compatible with mental arithmetic!

In his keynote address to the Industrial Statistics in Action 2000 Conference George Box compared the abundance of statistical applications in the literature of the social, medical and agricultural sciences with the paucity of examples in engineering, technology, and the physical sciences (Box 2000). If statistical computation was one of the obstacles to the advancement of statistical engineering, that has now been overcome with the advance of modern electronic computing technology.

In the 21st century computers are taken for granted, but it is interesting to digress for a few moments to look at the history of computing which was, and is still, intimately linked with the science of cryptography that today determines security on the internet. It will be seen that the pioneers of computing had their ups and downs.

The Victorian scientist Charles Babbage has been identified as the 'father of the computer'. According to Singh (2000) he 'pursued the life of a roving scholar, applying his mind to whatever tickled his fancy.' His credits include the invention of the speedometer and the cowcatcher, and the realisation that past climates could be studied by measuring the rings on ancient trees. He was also interested in statistics. As well as being an eccentric genius Babbage had endearing human characteristics. He had been interested in cyphers as a child and as a young man he was known to have investigated the practice of young Victorian lovers who sent encrypted messages to each other via the personal columns of newspapers. His reputation as a cryptanalyst led him to work on different occasions with a biographer, with a historian and with a barrister.

It was the proliferation of human errors in the mathematical tables of the day that inspired him to start work in 1823 on the development of a mechanical computing engine with financial support from the British government. No easy task, given the limitations of mechanical engineering at that time.

After working for ten years on his first design he then introduced a new design, but the government lost patience and the second model was never built until, using 20th century technology, the Science Museum in London reconstructed part of it according to Babbage's original design. The concept of a 'computing engine' capable of solving mathematical problems as well as calculating tables must still be attributed to Babbage.

It was not until the Second World War that electronic computing appeared on the scene. A mathematician, Max Newmann, at the Government Code and Cypher School, Bletchley Park, England conceived the possibility of using electronics to assist the code-breakers who were dealing with intercepted Lorenz-encrypted messages from Hitler to his generals. The idea was shelved by the authorities who deemed it technically impossible. Fortunately Tommy Flowers, an engineer at the research centre of the Post Office, ignored their scepticism and after ten months work delivered a machine embodying 1500 thermionic valves for installation at Bletchley in 1943.

The machine was code-named Colossus, which was appropriate judging by the photograph on page 136 of Singh's book. It seemed to fill a room from wall to wall and floor to ceiling. The fact that Colossus was programmable, and that it predated ENIAC (Electronic Numerical Integrator and Calculator) at Pennsylvania in 1945, made it the precursor of the modern electronic computer. Now let us return to the subject of statistical computing.

In the first half of the 20th century statistical computers were mostly hardworking lady assistants with a good command of arithmetic, access to Barlow's table of squares, and all the time in the world to complete their tasks. It is said that an early attempt at weather forecasting was hampered by taking a week to calculate the forecast one day ahead of the data. In due course the ladies' work was facilitated by the advent of hand-driven rotary mechanical calculators which required a certain degree of skill in rotation of the drum and alignment of the carriage for performing sums of squares. By mid-century these had been overtaken by electromechanical desk machines which would perform sums of squares automatically. These, in their turn, were overtaken by primitive electronic desk calculators which displayed data on special types of gas-filled electronic valves, each one containing ten electrodes shaped to form the digits zero to nine.

The emergence of liquid crystal display and integrated circuit technology made possible hand-held calculators which were capable of dealing with bivariate as well as univariate data (mean, variance, standard deviation, correlation, regression) a wide range of trigonometric and mathematical functions, physical constants, metric conversions, Boolean logic, etc. However, their use was confined to relatively small sets of data because of the effort of concentrating on entering data correctly, one digit at a time.

In the meantime the development of large-scale electronic computers during the second half of the century brought statistical computing to the point where users in the 1970s requiring complex analyses of large data sets could take their data and instructions on punched cards to a central computer located in an academic establishment or elsewhere in industry, commerce or government. They would collect the results of their analyses printed out in hard copy a few hours later, or the next day, depending on the length of the queue (providing the disk drive hadn't crashed or some other disaster had not befallen the operators). If the task was one that required the intervention of the user at various stages it could stretch out for weeks.

Computer users had the option of writing their own programmes in a computer language (Algol, Basic, Cobol, Fortran, etc.) or using one of the standard software packages that were available. For statistical analysis an early version of SPSS was in use. This originated as a statistical package for the social sciences, but in its present form it is now universally applicable in all statistical fields.

Since the early 1980s all of that has been swept aside by the advent of personal computers (PCs). Large computers still exist and are still developing, but they serve the needs of management in large organisations for accounting, production scheduling, stock control, process control, tax collecting, etc. When necessary they are still accessible to individual users through networks of computer terminals, but for most practical purposes the PC is an ideal tool for the individual.

The major breakthrough is that the free-standing PC on one's desk provides a service that is not only quicker (almost instantaneous) and more convenient, but in some respects more powerful. Information displayed on a video screen makes it possible for the user to intervene and control the analysis as it proceeds. Moreover, available PC software packages are more versatile and comprehensive than their predecessors, particularly in the matter of graphic displays. One need only compare the SPSS software of today with the best that was on offer 20 years ago. Modern experience shows that computing time is now usually only a small part of the time spent on a project. In the 1970s the time taken to access a computer was often a significant obstacle to progress, but not now.

A wide range of statistical software is now available for engineers and statisticians to use on their personal computers. *Microsoft Excel* is by far the most widely used spreadsheet software. Many engineers use it as a repository for their data and as an easy way to link different pieces of software together – they can run one analytical engineering programme and send the results to Excel, then read the results into another program. A range of statistical methods is available in the standard version of Excel.

Other general-purpose statistical packages include *SPSS, SAS, Stata, S-plus* and *Minitab*. *Minitab* is popular with consulting/training firms engaged in setting up Six Sigma quality improvement programmes. They use Minitab in their training of 'black belts' and 'green belts', and their clients buy Minitab licences for subsequent use. Specialised statistical packages that focus on design and analysis of experiments include *Design-Expert, ECHIP* and *JMP*. *Matlab* and *Mathcad* provide a wide range of mathematical functions useful for engineers. Both have built-in statistical functions. *Matlab* covers more ground because it has a Statistics 'toolbox' that can be bought separately from the main module.

However, there is always a price to pay for enhanced benefits. Users of PCs no longer enjoy the service of specialist computer operators to process their data. Instead, they must acquire familiarity with extensive interface mnemonics and commands before any data can be processed. Where possible, engineers should seek advice from statisticians on the most suitable choice of statistical software. If this is not possible they can refer to *Modern Industrial Statistics* (Kennet and Zacks 1998).

The rate of technology change seems to be accelerating, not diminishing. Computing equipment purchased today is obsolescent tomorrow. The printed word 'modern' is out of date almost before the ink is dry. Where does that leave those readers of this book who are faced with the daunting task of choosing the right PC from among so many that are strongly promoted in the market place? Or, again, those who already have a PC on their desk as a piece of office furniture and who wish to know what it can do for them as they get to grips with statistical engineering? What are the pros and cons?

Fortunately the growth of computer technology has spawned a host of publications designed to help those who, for one reason or another, need to get to grips with PCs, but are baffled by the variety of hardware and software, to say nothing of special terminology.

Readers who just want the basic facts about using a PC without a lot of technical jargon would do well to start with *10 Minute Guide to PC Computing* (O'Hara 1997). The author claims, 'Master the skills you need in 10 minutes or less.' That is not to be taken to mean that the whole field can be covered in ten minutes. There are 22 chapters in 146 pages, each one ending with a 10-minute exercise. The topics range from 'what is a computer' right through hardware, software, memory, disk drives, etc., up to the internet and e-mail. The readers must identify the particular field in which they wish to operate and the skill which they wish to acquire. The book is very readable, dealing with basic facts and using the minimum of technical jargon.

Another useful general guide is *The Which? Guide to Computers* (Wentk 2000) which is written for a wide readership, covering everything from offices to domestic use and children's games. It does not address the specialist issue of statistical computing. Nevertheless it makes good reading for a statistical engineering beginner. A seven-page section on health risks includes a comprehensive checklist. For example, the importance of good posture and regular exercise is stressed.

Engineers aiming to get to grips with specialist terminology right from the start might prefer *Buying a Personal Computer* (Brown 1999). The text ends with a glossary of over 50 technical terms without which much of the content of the preceding seven chapters would be incomprehensible to the uninformed reader. These chapters are full of good practical advice, beginning with thinking about what one actually needs, using a checklist of 50 possible requirements, before making a purchase. Payment by credit card rather than cash is recommended. A chapter devoted to understanding a PC's components gives detailed technical information on the role of individual components on the motherboard (i.e. the main printed circuit board). These are, the chipset which comprises one or more integrated circuits that handle data flow between the processor and input/output devices, the basic input/output system (BIOS), the CMOS RAM which draws its power from the battery on the motherboard and whose function is to store information that determines how the PC is configured, and the central processing unit (CPU). The technical aspects of random access memory (RAM) into which software is loaded, hard disk drives, floppy disk drives, compact disk drives with read only memory (CD-ROM), monitors, expansion cards, keyboard and mouse are discussed in sufficient detail to clarify the meaning of specifications found in advertisements for PCs. The chapter ends by warning purchasers that a depreciation of more than 50% in two years should be taken into account when planning the budget. A later chapter discusses anti-virus software and anti-theft software as well as providing a review of the variety of peripherals that can be added to a PC.

Budding statistical engineers have a lot in common with research students using computers in other disciplines. Good advice can be found in *Research Methods – Guidance for Postgraduates* (Greenfield 1996). In the chapter *Data handling on computers* Reese reminds us that the computer is only a mechanical slave. Unlike a human assistant, the computer does what it's told, not what you would like or expect. The user must remain in control of the method of analysis and the interpretation of the output. In other words, statistical engineers must understand the statistical process they are using as well as knowing the right commands to activate the software. He discusses the difference between PCs and Macs. PCs are computers that run the same programs as the IBM PC though they may not be made by IBM. Macs are made by the Apple Corporation. Disks and data files are not interchangeable between PCs and Macs. Statistical engineers should take note of his comment that spreadsheets which were introduced to serve accountants become cumbersome and error-prone when used as general programming tools and are not a replacement for custom-written statistical packages.

In a second chapter *Buying your own computer* Reese outlines questions that should be asked in the pre-purchase research phase. These apply to software as well as to hardware. In an example he gives technical details of the specification for a medium-priced system. Types of software are listed for word processing, text layout, spreadsheet, presentation graphics, data collection, statistical analysis, database programming, bibliographic database, networking and system maintenance. The list for statistical analysis suggests SPSS, SAS, Stata, Splus and Minitab as software in widespread use. A good variety of software is discussed and reviewed on the web site ProGAMMA (www.gamma.rug.nl).

In the second edition of *Research Methods – Guidance for Postgraduates* Greenfield (2002) Reese discusses recent developments in information technology, including compact disks, networks, digital scanners, laptop and palmtop computers, electronic mail, the Internet and the Web, which are having an impact on the PC world.

A useful spreadsheet for fitting distributions to data sets is available for general use, by request to the authors, with accompanying user instructions (Linsley *et al.* 2002, Industrial Statistics Research Unit, University of Newcastle upon Tyne.) This deals with Normal, Log-Normal, Half Normal, Extreme Value, Cauchy, Gamma, Logistic, Uniform, Chi-square, Exponential and Weibull distributions. This also has an Excel add-in *Essential Regression and Essential Experimental Design* which is available free from http://www.jowerner.homepage.t-online.de:80/download.htm.

Bibliography

Box, G. 2000: Statistics for discovery. In Coleman, S., Stewardson, D. and Fairbairn, L. *Proc. Industrial Statistics in Action 2000 Conference*. vol. I. pp. 23–35. University of Newcastle upon Tyne.

Brown, A. 1999: *Buying a Personal Computer*. Oxford: How To Books Ltd.

Greenfield, T. 2002: *Research Methods – Guidance for Postgraduates*. London: Arnold.

Hamilton, L.C. 1993: *Statistics with Stata 3*. Belmont, California: Duxbury Press.

Kenett, R.S. and Zacks, S. 1998: *Modern Industrial Statistics*. Pacific Grove, California: Duxbury Press.

Linsley, M., Fouweather, T., McGeeney, D., Stewardson, D.J. and Coleman, S.Y. 2002: Better understanding of automotive manufacturing reliability data using a spreadsheet-based model fitting tool. In *Proc. International Conference on Statistics and Analytical Methods in Automotive Engineering*, pp 245–254. Institution of Mechanical Engineers, Bury St Edmunds and London.

Lynn, P. 2002: Principles of Sampling. In Greenfield, T. *Research Methods – Guidance for Postgraduates*, pp 127–136. London: Arnold.

O'Hara, S. 1997: *10 Minute Guide to PC Computing*. Indianapolis: Macmillan Computer Publishing.

Penney, A. 2002: Instrumentation for Experiments. In Greenfield T, *Research Methods – Guidance for Postgraduates*, pp 160–168. London: Arnold.

Polak, T.A. and Pande, C. 1999: *Engineering Measurements, Methods and Intrinsic Errors*. London and Bury St Edmunds: Professional Engineering Publishing.

Reese, R.A. 2002: Data Handling on Computers. In Greenfield, T. *Research Methods – Guidance for Postgraduates*, pp 52–59. London, Arnold.

Reese, R.A. 2002: Buying Your Own Computer. In Greenfield, T. *Research Methods – Guidance for Postgraduates*, pp 60–65. London: Arnold.

Singh, S. 2000: *The Science of Secrecy – The History of Codes and Codebreaking*. London: Fourth Estate.

Wentk, R. 2000: *The Which? Guide to Computers*. London: Which? Ltd.

7

Quality Management

In case the preceding chapters should give the wrong impression, that statistical methods alone will ensure quality, a chapter on quality management is desirable to restore balance. Statistical engineering methods are no more, and no less, than tools used by engineers in the context of quality management. The statistical and managerial principles of quality are both important. Either one without the other is of much less value than the combination.

The way in which statistical engineering can serve quality management will now be examined by considering the four elements of a quality management model used in earlier publications (Morrison 1984, 1985, 1989). These are *planning, organising, directing and controlling*. The relationship between these four elements is that planning provides a basis for organising which in turn sets the stage for directing and controlling. In a wider context these are recognisable as the general principles of operations management.

7.1 Quality Planning

In the management literature two reasons are given for underlining the importance of management planning (Sisk 1969): its primacy from the standpoint of position in the sequence of management functions, and its pervasiveness as an activity that affects the entire organisation. Both these considerations apply with full weight in statistical engineering as well as in quality planning. Any organisational structure that does not

Statistics for Engineers: an Introduction   S.J. Morrison
© 2009 John Wiley & Sons, Ltd

relate to clearly defined objectives and policies is unlikely to be effective, and the direction and control of such an organisation will be a fruitless task. The factors affecting product quality are so widespread that it is difficult to identify an area of management that is not involved in some way. They are to be found not only in the technical areas of design and production, but also in marketing, purchasing, personnel, finance, legal and secretarial, and indeed in every sector of company activity. The same is true of statistical engineering, as was demonstrated in the variety of papers presented at the *Statistics in action 2000 Conference*. Hence the term *Total Quality Management* (TQM).

It follows that quality planning must not be considered in isolation, but has to be approached in the context of overall management planning. For the purpose of this chapter it can be assumed that a hypothetical industrial company will already have created a corporate plan for survival and long-term profitable growth. At the strategic boardroom level of planning, the importance of quality will have emerged from a comparative assessment of the corporate strengths and weaknesses of the company and its principal competitors, and the pursuit of quality improvement will have been recognised as one of the most important company objectives. If that point has not already been reached an established company will probably have no future in competitive markets. Given that it has been reached, the company will be ready to develop a suite of detailed tactical plans, one of which should deal specifically with quality, and should make provision for the application of statistical engineering methods.

It might be argued that quality can safely be subsumed under something else – production most likely – but that would be unrealistic in view of the recognition that must be given to quality as a prime determinant of competitiveness. Equally, the quality assurance plan must not be regarded as a substitute for other important elements in a suite of tactical plans. The quality function must act as a catalyst between other operational functions, as well as exercising authority in its own right. Statistical engineering should help to secure that goal.

The details of a tentative quality assurance plan for a typical industrial company can now be set out in terms of objectives, policy and procedures.

(a) Quality objectives

The prime objective should be to achieve a high degree of customer satisfaction, with due regard to quality costs. In this connection 'quality' must include any aspect of the product or service of which the customer may, within reason, take a critical view. The customer is entitled to be critical of design quality if the product specification falls short of his requirements. Quality of conformance will be an issue if the product, as supplied, does not meet a specification. The critical statistical role of variance synthesis as a tool of engineering design has already been demonstrated in Chapter 4. Quality of performance may leave something to be desired. Higher precision and accuracy are increasingly being expected of many products. Reliability is important at all times, if the product is functional, but especially if it represents a large capital investment. Quality of service can be a sensitive issue, in human as well as technical terms. The impression given by individuals can do much good, or much harm, to the company image.

Stating the quality objective in terms of customer satisfaction brings all these issues into a single focus and serves as a constant reminder that the customer who has grounds for dissatisfaction on any of them is at liberty to seek better satisfaction with a competitor.

It would be naïve to assume that each manufacturer should strive to become the quality leader in its own industry. In particular cases there may be sound commercial reasons for settling for something less. There is a market for the Mini as well as for the Rolls, and the word 'quality' has different meanings for different people. What is important is that the objective should be well chosen, in relation to the company's present capabilities and future prospects; and that it should be clearly defined and widely promulgated, so that everyone within the company can work towards a common goal.

(b) Quality policy

In terms of good management, quality policy must be more than a statement of intent, it must be manifest in a specific course of action designed to achieve the quality objective. That should include the deployment of statistical engineering as a tool of quality management. Variability is at the heart of every quality problem. Statistical methods provide the only satisfactory way of measuring variability, analysing it, identifying the sources, and bringing these under control. Managers and engineers who are unfamiliar with statistical methods are not well equipped for the quality task. They should not have to rely on external consultants telling them what to do. The establishment of in-house statistical engineering should be an essential feature of quality policy.

Finance is an important issue in quality management. Some companies are more vague and less specific on the financial side of their quality operations than on the technical side. The quality function is sometimes regarded as a nonproductive area suitable for cuts and economies rather than an area in which productive investment can be made. Statistical engineering must show it can pay its way. That should not be difficult, with powerful statistical methods in the hands of competent engineers.

It is difficult to see how such an important function as quality assurance can be properly managed without adequate budgetary provision and financial control. The fact that no formal provision may be made in a company does not mean there are no quality costs – it simply means they are hidden and are not susceptible to management. One of the first decisions to be made in quality policy-making must be how much to spend on running quality assurance operations. It may be necessary, at the outset, to make some rather crude estimates of the magnitude of existing costs of failure, and of appraisal and prevention, to reach a global figure. Whatever figure is arrived at, some appropriate amount of working capital should be set aside immediately for the development and operation of a quality cost system. If money is to be found to develop quality management as a separate management function it should be expected to pay its way alongside other management functions, and it should be accorded the financial services necessary for good management.

Once the financial decision is taken, other policy-making decisions will follow. It will be necessary to decide what sort of quality assurance system to adopt, how it is to be organised, and how it will be staffed and equipped. Here again, the employment of statistical engineering methods must not be lost sight of. There may well be a conflict between what is desirable and what can be afforded. The resolution of such conflict may rest in phasing the development of the system over a period of time.

Finally, it must be said that the policy-making decisions are of such fundamental importance that some of the more important ones can only be taken at director level. Hopefully, the director concerned will be well informed on the value of statistical engineering. It is axiomatic that quality begins in the board room even before the product reaches the drawing board.

(c) Quality procedures

In modern quality assurance practice it is necessary to create a system of standard procedures to service the quality function, and to determine the way in which it interacts with other functions. These procedures are managerial as well as technological and they extend far beyond the traditional statistical aids to manufacture and inspection, such as control charts and sampling plans. Nowadays they afford scope for applying all the statistical methods described in earlier chapters for carrying the attack on variability right back through research and development to the earliest conceptual stage of product design. They tend to form a complex highly interconnected system, but the complexity can be clarified by relating each procedure to one or more of three characteristic dimensions of the system. These are: (i) the production line dimension, (ii) the product life cycle dimension, (iii) the management dimension. Each of these will now be expanded.

The production line dimension is self-evident, but it must be extrapolated beyond the physical boundaries of the plant into the market in which the company's products are being sold and into the suppliers' market from which the company is drawing its raw materials, components, or supplies.

The following issues need to be dealt with:

1. Current market requirements
2. Current design/market requirements gap
3. Product liability
4. Field failures
5. Quality of customer service
6. Commissioning tests
7. In-plant finished product tests
8. Component and subassembly tests
9. Acceptance tests and raw materials and supplies
10. Approval of suppliers
11. Purchasing standards and specifications

The product life cycle dimension relates to the modern 'total quality' concept of designing quality and reliability into the product, rather than attempting to 'bolt them on' afterwards. For this to be done the quality function must involve itself with the design and development of new or improved products, and not simply confine its attention to the current product range. It is particularly important that the technique of variance synthesis described in Chapter 4 should be applied at the earliest possible stage of product design so as to identify and control sources of variability before they hit the production line and generate problems with customers. Statistical tolerancing described in Section 4.3 is also important.

The issues that are involved include the following:

1. Future market requirements
2. Market opportunities

3. New product target specifications
4. New product design specifications
5. New product production specifications
6. Process capability studies
7. Product reliability analysis
8. Product safety and reliability studies
9. Product quality forecasts
10. New quality operations and techniques
11. Quality cost budgets for new products

In the management dimension the procedures are those which are necessary for the effective management of the quality system and the coordination of quality activities within all the various operational areas of the company. These deal with the following issues:

1. Field failures
2. In-plant failures
3. Supply problems
4. Service problems
5. Quality costs
6. Quality budgets
7. Quality reporting
8. Quality audits
9. Education and training

Item 9 – Education and training – is critical. It is essential that everyone involved in quality operations has some acquaintance with statistical engineering operations, even though they may not themselves be statistical specialists.

The design of a suitable quality system for an individual company must relate to the quality problems and requirements of that particular company, but the issues involved are commonly those outlined above. It is important that a quality manual should be prepared, giving succinct descriptions of every procedure and identifying the purpose of each and the system inter-connections. The quality manual will be for in-house use, but if it is well prepared it can form the basis for a further publicity document to be used for the promotion of the company's quality image in the market place as well as for securing accreditation in national or international standardising systems. The presence of statistical engineering could well be featured in the publicity campaign.

7.2 Quality Organisation

A structure of organisation is created by identifying *roles*, by assigning *responsibility*, by delegating *authority*, and by creating *accountability*. It will be convenient to consider first, the role of the quality assurance function within the company before deciding details of the final structure of organisation.

The network of procedures that has already been described and which is necessary for the full development of a comprehensive quality plan extends far beyond the confines of production management. Apart from production, the branches of management in which essential quality-related activities have been identified include marketing, sales, service, purchasing, design, finance, accounting, personnel, as well as legal and secretarial. Many of these activities properly belong in the departments in which they are located, and it would be foolish to transfer them or duplicate them in a separate quality assurance department. The dominant role of the quality assurance function is therefore that of coordination, to ensure that all those activities are brought within the framework of an integrated system dedicated to achieve that most important objective – customer satisfaction. The responsibility and the accountability for the individual tasks must still remain with the departments in which they are performed, but their responsibility will include that of collaborating with the quality function wherever that is located. It may be in a separate quality department, or it may rest on a nominated individual in a department with a different primary role. The quality function will, of course, have unique duties of its own to perform.

The quality assurance role is managerial in its essentials, which are to plan, organise, direct, and control all quality-related activities in the company. The basic task of a quality manager is that which is common to all of management. It is to create and maintain an environment in which individuals can work together to accomplish group goals (Koontz and O'Donnell 1974).

Given that quality assurance must be found a place in the management organisation chart, where and how can it be accommodated? The resolution of conflicting departmental quality interests will be an important part of quality management. It cannot be assumed that conflict of interest will be less likely than in any other aspect of company activity. It is in the best interest of the company that quality management should be impartial, and independent of other departmental interests. It follows that quality assurance should be a top-level management function, reporting directly to the board and enjoying equal status with other top-level management functions (production, design, sales, marketing, etc.)

The question of which particular director the quality function should report to will depend on the board structure of each company. In some companies a quality director might be appointed, but in the absence of such an appointment the director to whom the quality function should report should be one carrying broad responsibilities for general management.

In a small company it may be necessary for individuals in the management hierarchy to wear more than one 'hat'. It is possible that the individual wearing the quality hat may have to carry other responsibilities as well. In such a case the identity of the quality function must be preserved by making a clear definition of the various distinctive functional roles.

Turning now to the responsibility and the accountability of the quality assurance function, the duties of a quality manager can be enumerated as follows:

1. To manage the quality function in such a way the company's quality objective of customer satisfaction is achieved.
2. To accomplish (1) with due regard to quality costs.
3. To develop and maintain the company's quality system.
4. To encourage employee motivation by whatever means may be appropriate (e.g. quality circles).
5. To be responsible for issuing quality specifications.
6. To take charge of metrology.
7. To take charge of inspection and testing.
8. To take charge of quality engineering and to ensure the effective solution of outstanding quality problems.
9. To plan and implement the quality elements in the new product development programme.
10. To act in an advisory capacity in the fields of standards, statistical methods, quality control and any other quality-related field.
11. To organise quality training for personnel in any part of the company.
12. To monitor the company's quality performance and quality costs.
13. To report on the discharge of the above responsibilities.

There remains the question of the internal organisation of the quality function. Given that the quality manager is a key executive, what sort and size of department will be required for support? Will the manager be expected to discharge all the quality duties personally, or will there be a staff of subordinates and personnel for delegation?

The quality assurance personnel can be divided into two groups. There must be an operating group engaged in measuring, testing and inspecting, and their numbers will be dictated by the scale of production and the requirements of the product specification. There may also be a managerial/ technical group which will include people engaged in solving quality problems, the clerical staff necessary for the maintenance of records, and the administrative staff of the department.

The key to the size of a quality department should be that small is beautiful. The quality manager will operate under the same regime as other operational managers, and must be prepared at all times to justify the existence of the group. There will be an operating budget at the disposal of the manager who will have to give a strict account of expenditure. The quality function must be seen to pay its way in terms of productivity and customer satisfaction.

7.3 Directing the Quality Function

At a time when a manufacturer stands in need of a turnaround in its fortune, the appointment of a quality manager to lead and direct quality operations is probably one of the most critical steps that can be taken. To face international competition in world markets that are becoming ever more quality conscious, manufacturers have to match their domestic as well as their foreign competitors with the best in quality management.

In the late 1970s the performance of the British economy was so outstandingly bad it came to be known as 'the British disease'. The management of the national economy was, of course, a matter for the government, but it was true to say that the cure required positive action at the level of the individual factory (Chatterton and Leonard 1979). It is interesting to note that among the six prime ingredients of the prescription recommended by these authors for curing the British disease there were four which were identified with good quality management, namely leadership, lateral communication, representative project groups, and job satisfaction.

(a) Leadership

What sort of an individual is required to administer the medicine? What are the criteria for selecting a quality manager?

A certain minimum level of technical competence is required, but the personal qualities are of great importance. The point was neatly put in pen-portraits of two types of professional managers (Duerr 1971): 'Mr Beta is the man with a business school background who, at the drop of a hat, will quote Herzberg, Maslow, McGregor, Likert, and tell you where Taylor and

Urwick went wrong. Mr Able can read a balance sheet, assess a marketing campaign, has a shrewd idea of what he can get out of a computer, but above all he knows people and can handle them.'

Quality manager Able has to be, above all, a leader with special qualities of leadership. The task of running a small department will be quite a minor one. The real leadership challenge will come in projecting the quality ethos into all other departments in which the principal quality tasks are carried out. That is where interpersonal skills will be tested to the limit.

The style of quality leadership must be both authoritative and participative – the dichotomy between these characteristics that crept into social science thinking was a nonsense. Authority will stem from professional commitment to quality and from the strength of will displayed in not allowing the quality operation to be turned away from its objective. At the same time participation must be inspired in all quarters and at all levels. The quality manager can accomplish nothing alone, but is totally dependent on others for the success of the quality programme.

Finally, the quality manager must have the ability to do all of these things without getting up other people's noses. As Duerr said, the manager must know people and be able to handle them.

(b) Lateral communication

Quality assurance depends more than most other industrial operations on lateral communication. It is a simplistic view, but nonetheless true, that product quality begins 'at the top' in the board room. But if every quality problem had to be referred upwards to the chief executive most of them would never get solved and a lot of other company business would suffer. The prime responsibility of the board is to create a management structure within which the quality department can work with other departments across the departmental boundaries over the whole span of the company organisation.

Collateral relationships between the quality department and other departments involve actions which must operate in both directions. The quality department provides services to other departments, but at the same time it has responsibilities for taking initiatives and promoting activities in these departments and for monitoring their activities. It follows that a two-way system of lateral communication is essential. The effectiveness of the communication system can be judged by the extent to which it promotes the acceptance of responsibility and commitment to quality at all levels in other departments.

The need to recognise the importance of informal as well as formal communication is of paramount importance. At any point in the system, be it in the design office, the production shop, or in the market place, there will be some individual who knows more than anyone else about an issue that is of vital importance. Such information is not always captured effectively in a formal system of communication, yet it may be essential to the solution of a problem. To use a mining analogy, not all of the coal-face problems can be solved on the surface back at headquarters.

(c) Project groups

As an extension of the general principle of not involving the chief executive in every minute quality detail, the quality manager and his opposite numbers in other departments should encourage their subordinates to use their initiative and to work together across departmental boundaries without referring every trivial detail to their departmental heads. To promote this, it may be useful to set up project groups to deal with important issues, with group members drawn from several departments and charged with the responsibility of working together to solve specific problems and then reporting back to management.

It is important that project groups should not become self-perpetuating oligarchies, soaking away precious manpower for all time. Each project group should have temporary terms of reference and should be disbanded when its task is complete.

It should be noted that a project group, as defined above, is a formal, (though temporary) extension of the management structure. As such, it must not be confused with a quality circle. The project group leader will usually be a manager, membership will be mandatory, the task will be delegated by management, and the project group will be owned by the company. The essential characteristics of quality circles differ from project groups (Robinson 1982). Quality circle leaders are first-line supervisors, membership is voluntary, the circle addresses things that go wrong at the members' own work place, and the members 'own' their circle.

Quality circles should never be seen as an alternative to a full-blooded quality management programme. To encourage quality circles to develop without a quality management framework is likely to promote a 'them' and 'us' atmosphere – the very thing that quality circles are intended to avoid. To encourage quality circles to develop alongside good quality management can be beneficial to both.

(d) Job satisfaction

Improved job satisfaction can come both from job enlargement and from job enrichment, and good quality management can contribute to each of these. One of the basic steps in modern quality assurance is to enlarge the production operators' jobs by placing on them the responsibility for checking their own work. Job enrichment comes from every member of the workforce being given a more active part to play in a company-wide quality programme. Quality is everybody's business. The routine production of a high-quality reliable product has to be seen in terms of human achievement, and there is no denying the satisfaction which that entails.

Certain industries have always been repositories of craft skills of the highest order. Modern quality assurance practice does not render traditional skills obsolete, but turns them to better use. With good quality management it should not be difficult for individuals to perceive that their personal goals are in harmony with company quality objectives.

7.4 Controlling the Quality Function

The control of quality operations, as distinct from the control of product quality, completes the circle of quality management. In simple terms, management control involves comparing the actual outcome of events with planned outcome, in order that intervention can be made where necessary. Sometimes corrective action will be needed to drive the actual outcome closer to what was planned, sometimes it will be recognised that the plan was defective and should be modified. Management control is an essential ongoing part of quality management, to ensure that the quality function is maintained on course and does not get deflected away from its objective.

It is the responsibility of the quality manager to design and implement a quality management control system as an extension of the quality planning function. The planning system will set the targets, and the control system will monitor progress towards targets.

It is desirable that the quality information system needed for control should form part of an overall management information system. For this to be so, an input of quality records and production records will be necessary, for which the quality manager and the production manager must be responsible. Because quality is everybody's business, these data should be accessible to all departments through a centralised management information system. Moreover, quality costs are a prime necessity for the quality manager and he should be drawing his quality cost data from the same source as all other departments, so that there will be no inconsistencies. It will, of course, be a responsibility of cost accounting to provide the input of cost data, but there must be close liaison between the accountant and the quality manager to ensure that the cost data will be realistic and in a format suitable for quality cost management.

It must be recognised that, of all the quality data available to managers for control purposes, cost data are inherently the most imprecise. Unlike the physical quality characteristics of the product which can be measured in absolute terms, quality costs (like all other costs) are abstract data based on certain conventions and assumptions. Whether these are suitable for the task in hand must be a matter for debate between the quality manager and the cost accountant. Some pitfalls in quality costing have been identified (Daisley *et al.* 1985).

7.5 Statistical Engineering

In this comprehensive review of quality management statistical engineering has been mentioned only occasionally. This does not mean that the statistical element of a quality system is of little or no importance compared with the managerial element. Statisticians and engineers must guard against the danger of managers and others becoming totally absorbed in complex management issues and failing to appreciate the value of statistical methods. The point has been made elsewhere that quality management without statistics is like a rowing boat without oars – other boats can go further and faster (Morrison, 1999).

The important conclusion to be drawn from this chapter is that there is a great deal more to statistical engineering than just entering numbers into formulae and pushing them through computers. To be effective the application of statistical methods in a quality system requires many other skills besides number crunching, and these have been indicated throughout this chapter.

The question must now be addressed, who should be responsible for conducting the statistical operations within a quality management system?

If a statistician is available the individual should be an applied statistician rather than a theoretician – someone who is prepared to come to terms with the inherent (and often necessary) pragmatism of engineers and managers. Compromise is necessary in real life industrial situations. A good approximation which yields a simple workable solution can be of more value than an exact formula of great complexity which is difficult to apply. On the other side, for a statistician's work to be successful it is essential that engineering colleagues develop an understanding of statistical concepts and methods.

There is, however, no good reason why the statistical task should not be undertaken by engineers themselves, using straightforward basic statistical methods described in earlier chapters, and calling for the services of a statistician when specialist advice is needed. It can be noted that certain techniques that are now regarded as essential items in the statistical toolkit were developed by individuals whose primary discipline was not statistics. Walter Shewhart (control charts) was a physicist. Harold Dodge (acceptance sampling inspection) was an engineer. Genichi Taguchi (experimental design) is an engineer. Present-day engineers should be encouraged to follow their example. There is unlimited scope for future development.

Whoever does take on board the task of applying statistical methods must be prepared to accept the management responsibilities outlined in this chapter, in particular those that involve communication with other personnel. The ability to translate statistical terminology into conventional English language and to explain the working of statistical methods in terms that can be understood by managers and others not trained in the subject is of vital importance.

One final warning to the budding statistical engineer: to avoid misleading others never base one's judgement solely on the level of significance in a statistical analysis. Always look for logical causation before announcing a result. Hunter (1981) cites the example of a recorded statistical correlation between size of population and number of storks, but anyone who draws the incorrect causal conclusion that storks bring babies and proceeds to shoot storks in the hope of reducing population will be disappointed. Statistical methods are powerful, but not infallible!

Bibliography

Chatterton, A. and Leonard, R. 1979: *How to avoid the British disease*. London: Mechanical Engineering Publications Ltd.

Daisley, P.A., Plunkett, J.J. and Dale, B.G. 1985: Quality costing in the UK, *Proc. World Quality Congress* 1984, 1, 557–567. London: Institute of Quality Assurance.

Duerr, C. 1971: *Management Kinetics: Carl Duerr on Communication*, p. 41. New York: McGraw-Hill.

Hunter, W.G. 1981: Six statistical tales, *The Statistician*, **30**(2), 107–118. Bury St Edmunds: The Institute of Statisticians.

Koontz, H. and O'Donnell, C. 1974: *Essentials of Management*. New York: McGraw-Hill.

Morrison, S.J. 1984: *Quality Assurance in the Machine Tool Industry*. University of Hull, Department of Engineering Design and Manufacture.

Morrison, S.J. 1985: Quality Management, *Proc. Institution of Mechanical Engineers*. **199**(B3), 153–159.

Morrison, S.J. 1989: Quality Assurance. In Lock, D. *Handbook of Engineering Management*, pp 575–594. Oxford: Heinemann Newnes.

Morrison, S.J. 1999: Total quality mismanagement. *Engineering Management Journal* **9**(6), 277–282. London: Institution of Electrical Engineers.

Robinson, M. 1982: *Quality circles, a practical guide*. Aldershot: Gower.

Sisk, H.L. 1969: *Principles of management*. Cincinatti: South Western Publishing Co.

8

Conclusion

The most important conclusion to be drawn from this text is that quality problems which plague production and embarrass customers can be identified, studied, and resolved, even before the first prototype of a new product has been created. This can be done simply by applying a straightforward statistical technique (i.e. variance synthesis) in engineering design.

If lead assessors who are responsible for examining quality management systems would insist on evidence that this is being done before they issued a certificate, a relatively ineffective bureaucracy would be transformed into a major factor in the promotion of quality. As things stand at the moment, too much attention is focused on documentation of quality management systems, too little on the nitty-gritty of quality.

In all of this there is a special role for academic engineers to play. It is their responsibility to raise new generations of graduates who are well versed in statistical engineering skills. The application of statistical methods, particularly in engineering design, should be an integral part of the engineering curriculum. It is up to authorities in the engineering profession who are responsible for accreditation of degree courses to see that this comes about.

Statistics for Engineers: an Introduction   S.J. Morrison
© 2009 John Wiley & Sons, Ltd

Appendix A

Guidelines

References [vol.I or II, pp.x–y.] are to items in Coleman, Stewardson and Fairbairn (2000), *Proc. Industrial Statistics in Action 2000 Conference.*

(a) Statistical engineering techniques

Automatic inventory control. **I**, 203–216.
Availability analysis. **I**, 63–73.
Bayes, **I**, 5–11. **I**, 328.
Beta distribution. **I**, 178–189.
Coefficient of variation. **I**, 153–162.
Control charts. **I**, 74–92, **II**, 27–41, **II**, 159, **II**, 175–188, **II**, 264–279.
Cusum. **I**, 121–137, **II**, 198–210.
Design of experiments. **I**, 5–11, **I**, 23–35, **I**, 113–120, **I**, 171–177, **I**, 265–273, **I**, 290–303, **II**, 43–51, **II**, 52–58, **II**, 56–69, **II**, 98–112, **II**, 113–125.
Evolutionary operation. **I**, 217–230.
Excess financial flow. **I**, 276–279.
Exponentially weighted moving average. **I**, 190–202.
Genetic algorithms. **I**, 217–230.
Markov models. **I**, 163–170.
Measurement. **I**, 326–327. **I**, 352–368.
Mentoring. **I**, 231. **I**, 244–249.
Modified simplex. **I**, 217–230.
Multivariate analysis of variance. **I**, 250–264.

(b) Areas of application of statistics

Appendix B

Recommended Books

Afifi, A.A. and Azen, S.P. 1979: *Statistical Analysis, A Computer Oriented Approach*, London: Academic Press.

Afifi, A.A. and Clark, V. 1996: *Computer-aided Multivariate Analysis*. London: Chapman & Hall.

Andrews, D.F. and Stafford, J.E. 2000: *Symbolic Computation for Statistical Inference*. Oxford University Press.

Arts. E. and Korst, J. 1989: *Simulated Annealing and Boltzman Machines*. Chichester: Wiley.

Baker, K.R. 1974: *Introduction to Sequencing and Scheduling*. New York: Wiley.

Barnett, V. 1991: *Sample Survey Principles and Methods*. London: Arnold.

Basford, K.E. and Tukey, J.W. 1999: *Graphical Analysis of Multi-response Data*. London: Chapman & Hall.

Beer, S. 1981: *Brain of the Firm*. Chichester: Wiley.

Bentley, J.P. 1999: *Introduction to Reliability and Quality Engineering*. Reading, MA:Addison-Wesley.

Bissel, D. 1994: *Statistical Methods for SPC and TQM*. London: Chapman & Hall.

Blom, G. 1958: *Statistical estimates and Transformed Beta Variables*. Reading, MA: Addison-Wesley.

Booker, J.D., Raines, M. and Swift, K.G. 2001: *Designing Capable and Reliable Products*. London: Butterworth-Heinemann.

Bossert, J.L. 1991: *Quality Function Deployment*: A Practitioner's Approach. Milwaukee, WI: ASQ Quality Press.

Bowker, A. and Lieberman, G. 1972: *Engineering Statistics*. Englewood Cliffs, NJ: Prentice-Hall.

Bowman, A. and McColl, J. 1999: *Statistics and Problem Solving*. London: Arnold.

Box, G.E.P. and Draper, N.R. 1969: *Evolutionary Operation – A Statistical Method for Process Improvement*. New York: Wiley.

Box, G.E.P. and Jenkins, G.M. 1976: *Time Series Analysis, Forecasting and Control*. San Francisco: Holden-Day.

Box, G.E.P., Hunter, W.G. and Hunter, J.S. 1978: *Statistics for Experimenters*. New York: Wiley.

Box, G.E.P. and Draper, N.R. 1987: *Empirical Model-Building and Response Surfaces*. New York: John Wiley.

Box, G.E.P., Jenkins, G.M. and Reinsel, G.C. 1994: *Time Series Analysis: Forecasting and Control* (3rd edn) Englewood Cliffs, NJ: Prentice- Hall.

Box, G.E.P. and Luceño, A. 1997: *Statistical Control by Monitoring and Feedback Adjustment*. New York: Wiley.

Brockwell, P.J. and Davis, R.A. 1996: *Introduction to Time Series and Forecasting*. New York: Springer-Verlag.

Brown, R.G. 1963: *Smoothing, Forecasting and Prediction of Discrete Time Series*. Englewood Cliffs, NJ: Prentice Hall.

Caplen, R.H. 1972: *A practical Approach to Reliability*. London: Business Books.

Carter, A.D.S. 1986: *Mechanical Reliability* (2nd edn) London: MacMillan Education.

Cassandras, C.G. 1993: *Discreet Event Systems: Modelling and Performance Analysis*. Homewood, IL: Irwin.

Caswell, F. 1995: *Success in Statistics*. London: John Murray.

Chatfield, C. 1995: *Problem Solving – A Statistician's Guide* (2nd edn) London: Chapman & Hall.

Chatfield, C. 1987: *The Analysis of Time Series – an Introduction*. London: Arrowsmith.

Chatfield, C. and Collins, A.J. 1980: *Introduction to Multivariate Analysis*. London: Chapman & Hall.

Checkland, P. 1981: *Systems Thinking, Systems Practice*. New York: Wiley.

Clarke, G.M. and Kempson, R.E. 1996: *Introduction to the Design and Analysis of Experiments*. London: Arnold.

Clarke, G.M. and Cooke, D. 1998: *A Basic Course in Statistics*. London: Arnold.

Cochrane, W.G. and Cox, G.M. 1957: *Experimental Designs*. New York: Wiley.

Cohen, S.S. 1985: *Operational Research*. London: Arnold.

Coleman, S., Greenfield, T., Jones, R., Morris, C. and Puzey, I. 1996: *The Pocket Statistician*. London: Arnold.

Condra, L.W. 1995: *Value Added Management by Design of Experiments*. London, Chapman & Hall.

Conover, W.J. 1999: *Practical Nonparametric Statistics*. New York: Wiley.

Cornell, J.A. 1990: *Experiments with Mixtures*. New York: Wiley.

Cox, D. and Reid, N. 2000: *The Theory of the Design of Experiments*. London: Chapman & Hall.

Cox, T.F. and Cox, M.A.A. 2000: Multidimensional Scaling (2nd edn) London: Chapman & Hall.

Crocker, D.C. 1990: *How to Use Regression Analysis in Quality Control*. Milwaukee, WI: ASQ Quality Press.

Daniel, C. 1976: *Application of Statistics to Industrial Experimentation*. New York: Wiley.

Davies, O.L. (ed.) 1954 *Design and Analysis of Industrial Experiments*. London and Edinburgh: Oliver & Boyd.

Davies, O.L. and Goldsmith, P.L. (ed.) 1972: *Statistical Methods in Research and Production with special reference to the chemical industry*. London and Edinburgh: Oliver & Boyd.

Day, R.G. 1993: *Quality Function Deployment*: Linking a Company with its Customers. Milwaukee, WI: ASQ Quality Press.

DelVecchio, R.J. 1997: *Understanding Design of Experiments*: A Primer for Technologists. Milwaukee, WI: ASQ Quality Press.

Deming, W.E. 1986: *Out of the Crisis*. Cambridge, MA: MIT Press.

Deming, W.E. 1993: *The new Economics*. Cambridge, MA: MIT Press.

Devor, E., Chang, T. and Sutherland, J. 1992: *Statistical Quality Design and Control*. New York: Macmillan.

Dodge, H.F. and Romig, H.G. 1959: *Sampling Inspection Tables: Single and Double Sampling*. New York: Wiley.

Dovich, R.A. 1992: *Quality Engineering Statistics*. Milwaukee, WI: American Society for Quality.

Draper, N.R. and Smith, H. 1998: *Applied Regression Analysis* (3rd edn) New York: Wiley.

Eisenhart, C. Hastay, M.W. and Wallis, W.A. (eds) 1947: *Techniques of Statistical Analysis*. New York: McGraw Hill.

Fisher, R.A. 1960: *The Design of Experiments* (7th edn) Edinburgh: Oliver & Boyd.

Fowlkes, W.Y. and Creveling. C.M. 1995: *Engineering Methods for Robust Design*. Reading, MA: Addison-Wesley.

Freeman, H.A., Friedman, M., Mosteller, F. and Wallis, W.A. 1948: *Sampling Inspection: Principles, Procedures and Tables*. New York: McGraw-Hill.

Fu, M. and Hu, J.Q. 1997: *Conditional Monte Carlo: Gradient Estimation and Optimisation Applications*. Boston: Kluwer.

Glasserman, P. 1991: *Gradient Estimation via Perturbation Analysis*. Norwell, MA: Kluwer Academic Publisher.

Goldberg, D.E. 1989: *Genetic Algorithms in Search, Optimisation and Machine Learning*. Reading, MA: Addison-Wesley.

Goldratt, E. 1993: *The Goal*. London: Gower.

Gower, J.C. and Hand, D.J. 1996: *Biplots*. London: Chapman & Hall.

Grant, E.L. and Leavenworth, R.S. 1988: *Statistical quality control*. New York: Wiley.

Grove, D.M. and Davis, T.P. 1992: *Engineering Quality and Experimental Design*. London: Longman.

Gunst, R.F. and Mason, R.L. 1991: *How to Construct Factorial Experiments*. *Milwaukee*, WI: ASQ Quality Press.

Harry, J.H. 1988: *The Nature of Six Sigma Quality*, Rolling Meadows, IL: Motorola University Press.

Harry, M. and Schroeder, R. 2000: *Six Sigma: The Breakthrough Management Strategy Revolutionising the World's Top Companies*. New York: Doubleday.

Hartley, B. and Hawkes, T.O. 1970: *Rings, Modules and Linear Algebra*. London: Chapman & Hall.

Harvey, A.C. 1989: *Forecasting Structural Time Series Models and the Kalman Filter*. Cambridge: The University Press.

Hastie, T.J. and Tibshirani. R.J. 1990: *Generalised Additive Models*. Chapman & Hall.

Hawkins, D.M. & Olwell, D.H. 1998: *Cumulative Sum Charts and Charting for Quality Improvement*. New York: Springer Verlag.

Hearn, G.E. and Metcalf, A.V. 1995: *Spectral Analysis in Engineering: Concepts and Cases*. London: Arnold.

Hewitt, C.N. 1995: *Methods of Environmental Data Analysis*. London: Chapman & Hall.

Hines, W.H. and Montgomery, 1990: *Probability and Statistics in Engineering and Management Science* (4th edn) New York: Wiley.

Ho, Y.C. and Cao, X.R. 1991: *Perturbation Analysis of Discrete Event Dynamic Systems*. Norwell, MA: Kluwer Academic Publisher.

Hoaglin, D.C., Mosteller, F. and Tukey, J.W. 1983: *Understanding Robust and Explanatory Data Analysis*. New York: Wiley.

Holland, J.H. 1975: *Adaptation in Natural and Artificial Systems*. Ann Arbor: The University of Michigan Press.

Hurst, K. 1999: *Engineering Design Principles*. London: Edward Arnold.

Imaii, Masaaki. 1986: *Kaizen – The Key to Japan's competitive Success*. New York: Random House.

Imaii, Masaaki. 1998: *Gemba Kaizen*. New York: Random House.

Ishikawa K. (Transl. Loftus, J.H.) 1990: *Guide to Quality Control*. London: Chapman and Hall.

Jackson, J.E. 1991: *A Users Guide to Principal Components*. New York: Wiley.

Johnson, N.L. and Leone, F.C. 1964: *Statistics and experimental design*. New York: Wiley.

Johnson, N.L. and Kotz, S. 1970: *Distributions in Statistics, Continuous Univariate Distributions-2*. New York: Wiley.

Jolliffe, I.T. 1986: *Principal Component Analysis*. New York: Springer-Verlag.

Juran, J.M. (ed.) 1995: *A History of Managing for Quality*. Milwaukee, WI: ASQC Quality Press.

Juran, J.M. and Godfrey, A.B. 1999: *Juran's Quality Handbook*, 5th edn, Milwaukee, WI: ASQ Quality Press.

Juran, J.N. and Gryna, F.M. 1988: *Quality Planning and Analysis*. New York: McGraw-Hill.

Kaplan, R.S. and Norton, D.P. 1996: *The Balanced Scorecard*: Translating Strategy into Action. HBS Press.

Kendall, M.G. and Buckland, W.R. 1957: *A dictionary of statistical terms*. London and Edinburgh: Oliver & Boyd.

King, J.R. 1971: *Probability Charts for Decision Making*. New York: Industrial Press.

Kohonen, T. 1989: *Self-organisation and Associative Memory*. New York: Springer-Verlag.

Kunio Shirose. 1992: *TPM for Workshop Leaders*. USA: Productivity Press.

Laarhoven, Van P.J.M. and Arrts, E.H.L. 1987: *Simulated Annealing: Theory and Applications*. Dordrecht: Lancaster Reidel.

Lees, F.P. 1983: *Loss Prevention in the Process Industries*. London: Butterworth.

Lindley, D.V. and Miller, J.C.B. 1962: *Cambridge Elementary Statistical Tables*. Cambridge: The University Press.

Lindley, D.V. and Scott, W.F. 1995: *New Cambridge Statistical Tables*. Cambridge: The University Press.

Mackenzie, L.D. and Cornwell, D.A. 1998: *Introduction to Environmental Engineering* (3rd edn) New York: McGraw-Hill.

Makridakis, S., Wheelwright, S.C. and McGee, V.E. 1983: *Forecasting: Methods and Applications* (2nd edn). New York: Wiley.

Mandel, J. 1991: *Evaluation and Control of Measurements*. New York: Marcel Dekker.

Manly, B.J.F. 1994: *Multivariate Statistical Methods*: A Primer (2nd edn) London: Chapman & Hall.

Mardia, K.V., Kent, J.T. and Bibby, J.M. 1979: *Multivariate Analysis*. London: Academic Press.

Martens, H. and Næs, T. 1989: *Multivariate Calibration*. New York: Wiley.

Mason, R.L., Gunst, R.F. and Hess, J.L. 1989: *Statistical Design and Analysis of Experiments: With Applications in Engineering and Science*. New York, Wiley.

Merli, G. *Managing by Priority*. New York: Wiley.

Metcalfe, A.V. 1994: *Statistics in Engineering – A Practical Approach*. London: Chapman & Hall.

Metcalfe, A.V. 1997: *Statistics in Civil Engineering*. London: Arnold.

Metcalfe, A.V. 2000: *Statistics in Management Science*. London: Arnold.

Miller, A.J. 1990: *Subset Selection in Regression*. New York: Chapman & Hall.

Montgomery, D.C. 2001: *Design and Analysis of Experiments* (5th edn) New York: Wiley.

Montgomery, D.C. 1991: *Introduction to Statistical Quality Control* (6th edn) New York: Wiley.

Montgomery, D.C. 1991 *Statistical Process Control*. New York: Wiley.

Montgomery, D.C., Runger, D.C. and Hubele, F.N. 2007: *Engineering statistics* (4th edn) New York: Wiley.

Muller, B., Reinhardt, J. and Strickland, M.T. 1995: *Neural Networks: An Introduction*. New York: Springer-Verlag.

Murdoch, J. and Barnes, J.A. 1986: *Statistical Tables for Science, Engineering, Management and Business Studies*. London: Macmillan.

Myers, R.H. and Montgomery, D.C. 1995: *Response Surface Methodology: Process and Product Optimisation using Designed Experiments*. New York: Wiley.

Nakajima, S. 1988: *Introduction to TPM: Total Productive Management*. USA: Productivity Press.

Neave, H.R. 1990: *The Deming Dimension*. SPC Press.

Nelson, W. 1982: *Applied Life Data Analysis*. New York: Wiley.

Neslen, R. 1998: *Construction Maths Advanced*. London: Edward Arnold.

Oakland, J.S. 1996: *Statistical Process Control*. London: Butterworth-Heinemann.

Oakland, J.S. 1994: *Total Quality Management* (2nd edn) London: Butterworth-Heinemann.

Oakland, J.S. and Porter, L.J. 1994: *Cases in Total Quality Management*. London: Butterworth-Heinemann.

O'Connor, P.D.T. 1991: *Practical Reliability Engineering* (3rd edn) Chichester: Wiley.

O'Connor, P.D.T. 2001: *Test Engineering – A Concise Guide to Cost-effective Design, Development and Manufacture*. Chichester: Wiley.

Owen, M. 1990: *SPC and Continuous Improvement*. Bedford, UK: IFS Publications.

Pages, A. and Gondran, M. 1986: *System Reliability Evaluation and Prediction in Engineering*. London: North Oxford Academic.

Park, A. 1991: *Taguchi Methods: Introduction to Quality Engineering*. Dearborn, MI: American Supplier Institute.

Pearson, E.S. and Hartley, H.O. 1966: *Biometrica Tables for Statisticians*. Cambridge: The University Press.

Pennick, A.M. 1999: *Kempe's Engineers Year-book* 1999. London: Miller Freeman Publications.

Pentecost, A. 1999: *Analysing Environmental Data*. London: Pearson Education Ltd.

Phadke, M.S. 1989: *Quality Engineering Using Robust Design*. London: Prentice-Hall.

Polak, T.A. and Pande, C. 1999: *Engineering Measurements: Methods and Intrinsic Errors*. Bury St Edmunds: Professional Engineering Publishing Ltd.

Pole, A., West, M. and Harrison, P.J. 1997: *Bayesian Forecasting and Dynamic Models*. New York: Springer-Verlag.

Potarzycki, C. 1998. *Applied Maths for Engineering*. London: Arnold.

Press, W.H., Vetterling, W.T., Tenkolsky, S.A. and Flannery, B.P. 1992: *Numerical Recipes in Fortran. The Art of Scientific Computing*. Cambridge: The University Press.

Rasmussen, J. 1986: *Information Processing and Human–Machine-Interaction – An Approach to Cognitive Engineering*. Amsterdam: North Holland.

Rees, D.G. 2000: *Essential Statistics*. London: Chapman and Hall.

Ripley, B.D. 1996: *Pattern Recognition and Neural Networks*. Cambridge: The University Press.

Ross, S. 1997: *Simulation*. London: Academic Press.

Ross, P. 1996: *Taguchi Techniques for Quality Engineering* (2nd edn) New York: McGraw-Hill.

Rubinstein, R.Y. 1986: *Monte Carlo Optimisation, Simulation and Sensitivity of Queuing Networks*. New York: Wiley.

Rumelhart, D.E. and McClelland, J.L. 1986: *Parallel Distributed Processing: Explorations in the Microstructure of Cognition*. Cambridge, Mass: MIT Press.

Russell, S.J. and Norvig, P. 1995: *Artificial Intelligence: A Modern Approach*. Englewood Cliffs: Prentice Hall.

Ryan, T.P. 2000: *Statistical Methods for Quality Improvement*. Chichester: Wiley.

Schey, J.A. 1987: *Introduction to Manufacturing Processes* (2nd edn) New York: McGraw-Hall.

Schwefel, H-P. 1995: *Numerical Optimisation of Computer Models*. Chichester: Wiley.

Schwefel, H-P. 1995: *Evolution and Optimum Setting*. New York: Wiley.

Senge, P.M. 1990: *The Fifth Discipline: The Art and Practice of the Learning Organisation*. New York: Doubleday.

Senge, P.M. 1993: *The Fifth Discipline*. Random House Business Books.

Shewhart, W.A. 1931: *Economic Control of Quality of a Manufactured Product*. New York: Van Nostrand.

Strang, G. *Linear Algebra and its Applications* (3rd edn) San Diego: Harcourt Brace Javanovich.

Swift, K.G. and Field, S.W. 1996: *Effecting a Quality Change, an Engineering Approach*. London: Arnold.

Swift, K.G. and Booker, J.D. 1997: *Process Selection from Design to Manufacture*. London: Arnold.

Taguchi, G. and Wu, Y. 1985: *Introduction to Off-Line Quality Control*. Nagoya: Central Japan Quality Control Association.

Taguchi, G. 1986: *Introduction to Quality Engineering*. Tokio: Asian Productivity Association.

Taguchi, G. 1987: *System of Experimental Design*. New York: Unipub/Kraus International Publications.

Tiao, G.C.(ed.) 2000: *Box on quality and discovery*. New York: Wiley.

Tippet, L.H.C. 1952: *The Methods of Statistics*. London: Williams and Norgate.

Tokaturo Suzuki. 1994: *TPM in Process Industries*. USA: Productivity Press.

Vardeman, S.B. 1999: *Statistical Quality Assurance Methods for Engineers*. Milwaukee, WI: ASQ Quality Press.

Vining, G.G. 1998: *Statistical Methods for Engineers*. Duxbury.

Wadsworth, H.M., Stephens, K.S. and Godfrey, A.B. 1986: *Modern Methods for Quality Control and Improvement*. Singapore: Wiley.

Webb, A. 1999: *Statistical Pattern Recognition*. London: Edward Arnold.

West, M. and Harrison, P.J. 1997: *Bayesian Forecasting and Dynamic Models* (2nd edn) New York: Springer-Verlag.

Wetherill, G.B. and Brown, D.W. 1991: *Statistical Process Control*. London: Chapman and Hall.

Wheeler, D. and Chambers, D.S. 1992: *Understanding Statistical Process Control*. Knoxville: SPC Press.

Wheeler, D. 1993: *Understanding Variation – The Key to Managing Chaos*. SPC Press.

Whittle, P. 1963: *Prediction and Regulation*. London: English Universities Press.

Woodward. R.H. and Goldsmith, P.L. 1964: Cumulative sum techniques. (ICI Monograph No.3 in the 'mathematical and statistical techniques for industry' series.) Edinburgh: Oliver & Boyd.

Wu, C.J.F. and Hamada, M. 2000: *Experiments: Planning, Analysis, and Parameter Design Optimization*. New York: Wiley.

Wynn, H.P., Riccomagno, E. & Pistone, G. 2000: *Algebraic Statistics*. London: Chapman and Hall.

Appendix C

Periodicals

Advances in Applied Probability. (Applied Probability Trust).
Annals of Mathematical Statistics. (Institute of Mathematical Statistics).
Annals of Probability. (Institute of Mathematical Statistics).
Annals of Statistics. (Institute of Mathematical Statistics).
Australian and New Zealand Journal of Statistics. (Blackwell).
Biometrical Journal. (Wiley Interscience).
Biometrika. (Oxford University Press).
British Journal of Mathematical and Statistical Psychology. (British Psychological Society).
Canadian Journal of Statistics. (Statistical Society of Canada).
Computational Statistics. (Springer).
Computational Statistics and Data Analysis. (Elsevier Science).
Econometrica, Journal of the Econometric Society. (Econometric Society).
Engineering & Technology. (Institution of Engineering and Technology).
Environmetrics. (John Wiley & Sons).
Fuzzy Sets and Systems. (Elsevier Science).
Insurance: Mathematics and Economics. (Elsevier Science).
International Journal of Game Theory. (Springer).
International Statistical Review. (International Statistical Institute).
Journal of the American Statistical Association. (American Statistical Association).

Journal of Applied Probability. (Applied Probability Trust).

Journal of Applied Statistics. (Taylor & Francis).

Journal of Computational and Graphical Statistics. (American Statistical Association).

Journal of Multivariate Analysis. (Academic Press).

Journal of Quality Technology. (American Society for Quality).

Journal of the Royal Statistical Society: Series A (Statistics in Society). (Royal Statistical Society).

Journal of the Royal Statistical Society: Series B (Methodological). (Royal Statistical Society).

Journal of the Royal Statistical Society: Series C (Applied Statistics). (Royal Statistical Society).

Journal of the Royal Statistical Society: Series D (The Statistician). (Royal Statistical Society).

Journal of Statistical Computation and Simulation. (Gordon & Breach).

Journal of Statistical Planning and Inference. (Elsevier Science).

Journal of Theoretical Probability. (Kluwer Academic Publishers Group).

Management Today. (The Institute of Management).

Metrika: International Journal of Theoretical and Applied Statistics. (Springer).

Multivariate Behavioral Research. (Lawrence Erlbaum Associates).

Oxford Bulletin of Economics and Statistics. (Blackwell).

Probability in the Engineering and Informational Sciences. (Cambridge University Press).

Probability Theory and Related Fields. (Springer).

Professional Manager. (The Institute of Management).

Quality Engineering. (American Society for Quality).

Quality Management Journal. (American Society for Quality).

Quality and Quantity. (Kluwer Academic Publishers Group).

Quality World. (Institute of Quality Assurance).

Scandinavian Journal of Statistics. (Danish Society for Theoretical Statistics, the Finnish Statistical Society, the Norwegian Statistical Society and the Swedish Statistical Association).

Statistical Computation and Simulation. (Gordon & Breach).

Statistical Science. (Institute of Mathematical Statistics).

Statistics and Probability Letters. (Elsevier Science).

Stochastic Hydrology and Hydraulics. (Springer).

Stochastic Processes and their Applications. (Elsevier Science).

Technometrics. (American Society for Quality and American Statistical Association).

The TQM Magazine. (MCB University Press).

NOTE

The periodicals listed above cater for the quality-related technological managerial/statistical interest of the statistical engineer. Most, but not all, have a strong statistical interest. In some it is the primary interest (e.g. the journals of the Royal Statistical Society and the American Statistical Association). In others it is linked to a professional interest such as psychology. It is encouraging to detect a growing interest in statistics in the IET fortnightly flagship *Engineering & Technology*, but disappointing to note that the statistical interest appears to have diminished in *Quality World* published by the Institute of Quality Assurance. This, along with the very rare occurrence of statistical papers in *The TQM Magazine* reflects an excessive concentration on quality management and certification to the detriment of the statistical interest. It is not surprising, but neither is it pleasing, that statistical methods are neglected in the periodicals of the Institute of Management.

Appendix D

Supplementary Bibliography

Useful references not appearing already in chapter reference lists or in Appendix B are classified below.

(a) Business and industry

Box, G.E.P. and Bisgaard, S. 1987: The Scientific Context of Quality Improvement. *Quality Progress*, June, 54–61. American Society for Quality.

BS EN ISO 9001, 1994: *Model for Quality Assurance in Design, Development, Production, Installation and Servicing*. London: British Standards Institution.

Durham, K. 1989: *Productivity and Statistics. Phil. Trans. Roy. Soc. A* **327**, 481–485. London: Royal Society.

Greenfield, T. and Siday, S. 1980: Statistical computing for business and industry. *The Statistician* **29** (1), 33–35. Bury St Edmunds: The Institute of Statisticians,

Morrison, S.J. 1981: *A Comparative Quality Assurance Study of British, European, American and Japanese Manufactured Products*. University of Hull, Department of Operational Research. (Reprinted by Department of Engineering Design and Manufacture, 1985.)

Royal Society. 1989: *Industrial Quality and Productivity with Statistical Methods – A Joint Symposium of the Royal Society and the Royal Statistical Society. Phil. Trans. Roy. Soc. A* **327**, 1596, 477–638. London: Royal Society.

Statistics for Engineers: an Introduction   S.J. Morrison
© 2009 John Wiley & Sons, Ltd

Wheelwright, S.C. 1984: Strategic management of manufacturing. *Advances in Applied Business Strategy* **1**, 1–15. JAI Press Inc. USA.

(b) Control

Champ, W.S., Woodall, W.H. and Mohsen, H.A. 1991: A generalised quality control procedure. *Statistics and Probability Letters*, **11**, 211–218.

ISO 2859–1 1999: Sampling Schemes Indexed by AQL for Lot-by-lot Inspection. London: British Standards Institute.

Jones, R. and Winterbottom, A. 1997: Integrating statistical and engineering process control. *Quality World*, January, London: Institute of Quality Assurance.

Morrison, S.J. 1958: The lognormal distribution in quality control. *Applied Statistics* **VII**, **3**, 160–172. London: Royal Statistical Society.

Morrison, S.J. 1962: Target distributions for quality control. *Applied Statistics* **XI**, 1, 55–63. London: Royal Statistical Society.

Papanassios, A., Metcalfe, A. and Lowther, A. 1999: Eliptical control charts: a practical application. *Quality World*, October, 2–34. London: Institution of Quality Assurance.

(c) Engineering

Batchelor, R. and Swift, K.G. 1996: Conformability and Analysis in Support of Design for Quality. *Proc. Inst. Mech. Engrs*. **B210**, 37–47. London: Institution of Mechanical Engineers.

BS 6835 1988: *Determination of the Rate of Fatigue Crack Growth in Metallic Materials*. London: British Standards Institute.

Edwards, S.P. Grove, D.M. and Wynn, H.P. 1998: *Statistics for engine optimisation*. Bury St Edmunds: Professional Engineering Publishing.

Edwards, S.P. *et al.* 2002: International conference on statistics and analytical methods in automotive engineering. Bury St Edmunds: Professional Engineering Publishing.

Jebb, A. and Wynn, H.P. 1989: Robust Engineering Design post-Taguchi. *Phil. Trans. Roy. Soc.* **A 327**, 605–616. London: Royal Society.

Morrison, S.J. 1981: Importance of quality and reliability. *Proc. IEE*, **128A**, 7, 520–524. London: Institution of Electrical Engineers.

Morrison, S.J. 1982: The importance of quality and reliability as determinants of international competitiveness. *Proceedings of Quality in Electronics Conference, Raleigh, NC* pp 6–10. American Society for Quality Control.

Morrison, S.J. 2000: Statistical Engineering Design. *The TQM Magazine* **12**, 1, 26–30. Bradford: MCB University Press.

Morrison, S.J. 2001; Quality engineering design. *Manufacturing Engineer* **80**, 3, 110–112. London: Institution of Electrical Engineers.

Morrison, S.J. 2002: The missing Link, *Engineering Science and Education Journal*, **11** (4), 133–138. London: Institution of Electrical Engineers.

Swift, K.G. and Allen, A.J. 1994: Product Variability Risks and Robust Design. *Proc. Inst. Mech. Engrs.* **B208** (B), 9–19.

Swift, K.G. and Booker, J.D. 1996: Engineering for Conformance. *The TQM Magazine* **8** (3), 54–60.

(d) Experiments

Anderson, M.J. and Whitcomb, P.J. 2000: How to design and analyse mixture designs that include process factors and/or categorical variables. In: Coleman, S.Y., Stewardson, D. and Fairbairn, L. 2000: *Industrial Statistics in Action*, **II**, 52–58. University of Newcastle upon Tyne.

Ankenman, B.E. 1999: Design of experiments with two- and four-level factors. *Journal of Quality Technology.* **V**, 31, 363–375.

Bisgaard, S. 1992: Industrial use of statistically designed experiments: case study references and some historical anecdotes. *Quality Engineering.* **4**, 547–562.

Box, G.E.P. and Wilson, K.B. 1951: On the experimental attainment of optimum conditions. *Jour. Roy. Stat. Soc.* **B**, 23, 1–45. London: Royal Statistical Society.

Box, G.E.P. and Hunter, J. 1961: The 2^{k-p} fractional factorial designs, parts 1&2. *Technometrics*, **3**, 311.

Box, G.E.P. and Meyer, R.D. 1986: Dispersion effects from fractional designs. *Technometrics*, **28**, 1, 19–27.

Box, G.E.P. and Jones, S. 1992: Split-plot designs for robust experimentation. *Journal of Applied Statistics.* **19** (1), 3–25.

Daniel, C. 1959: Use of Half-Normal Plot in Interpreting Factorial Two-Level Experiments. *Technometrics* **1**, 311–341.

Douglass, J. and Coleman, S.Y. 2000: Improving product yield using a statistically designed experiment. In: Coleman, S.Y., Stewardson, D. and Fairbairn, L. 2000: *Industrial Statistics in Action*, **I**, 171–177. University of Newcastle upon Tyne.

Goh, T.N. 2000: A pragmatic approach to experimental design in industry. In: Coleman, S.Y., Stewardson, D. and Fairbairn, L. 2000: *Industrial Statistics in Action*, **II**, 43–51. University of Newcastle upon Tyne. Also in *Journal of Applied Statistics*, 2001, **28**, 3 and 4, 391–398.

Joiner, B.L. and Campbell, C. 1976: Designing experiments when run order is important. *Technometrics* **18** (3), 249–259.

Martin, R.J., Platts, L.M., Seddon, A.B. and Stillman, E.C. 2000: The design and analysis of a mixture experiment on glass durability. In: Coleman, S.Y., Stewardson, D. and Fairbairn, L. 2000: *Industrial Statistics in Action*, I, 265–275. University of Newcastle upon Tyne.

Plackett, R.I. and Burman, J.P. 1946: The design of optimum multifactorial experiments. *Biometrika*. **33**, 305–325.

Porter, D. 2000: The application of DOE techniques to the development of a novel pressure swing absorber. In: Coleman, S.Y., Stewardson, D. and Fairbairn, L. 2000: *Industrial Statistics in Action*, I, 113–120. University of Newcastle upon Tyne.

Stewardson, D.J., Drewett, L., Silva, L., Budano, S., Joller, A., Mertens, J and Baudry, G. 2000: Using designed experiments and the analysis of statistical error to determine change points in fatigue crack growth rates. In: Coleman, S.Y., Stewardson, D. and Fairbairn, L. 2000: *Industrial Statistics in Action*, II, 59–69. University of Newcastle upon Tyne.

Stewardson, D.J., Porter, D. and Kelly, T. 2000: The danger posed by saddle points and other problems when using central composite designs. In: Coleman, S.Y., Stewardson, D. and Fairbairn, L. 2000: *Industrial Statistics in Action*, I, 280–289. University of Newcastle upon Tyne.

(e) Historical

Box, J.F. 1978: *R.A. Fisher: The Life of a Scientist*. New York: John Wiley.

Field, A. 1998: Quality heroes – W Edwards Deming (1900–1993). *Quality World*, March, 26–27. London: Institute of Quality Assurance.

Field, A. 1998: Quality heroes – Joseph Juran (born 1904). *Quality World*, April, 36–37. London: Institute of Quality Assurance.

Field, A. 1998: Quality heroes – Armand V Feigenbaum. *Quality World*, May, 24–26. London: Institute of Quality Assurance.

Field, A. 1998: Quality heroes – Dr Kaoru Ishikawa (1915–1989). *Quality World*, June, 36–37. London: Institute of Quality Assurance.

Hunter, J.S. 1999: Statistical process control: a glimpse at its past and future. *Quality Progress*, December 54–57. Milwaukee, WI: American Society for Quality.

Juran, J.M. 1995: *A History of Managing for Quality*. Milwaukee, WI: ASQC Quality Press.

Morrison, S.J. 2002: The missing Link, *Engineering Science and Education Journal*, II, 4, 133–138. London: Institution of Electrical Engineers.

Noguchi, J. 1995: The legacy of W. Edwards Deming. *Quality Progress*, December 35–37. Milwaukee, WI: American Society for Quality.

Nonaka, I. 1995: The recent history of managing for quality in Japan, in Juran, *A History of Managing for Quality*, 517–552. Milwaukee, WI: ASQC Quality Press.

Pearson, E.S. 1935: Discussion of Mr Tippet's paper. *Journal of the Royal Statistical Society, Supplement* **2**, 57–58.

(f) Six sigma

Bendell, T. 2000: What is six sigma? *Quality World*, January, 14–17. London: Institute of Quality Assurance.

Blakesley, J.A. 1999: Implementing the six sigma solution. *Quality Progress*, July 77–85. Milwaukee, WI: American Society for Quality.

Snee, R.D., 1999: Why should statisticians pay attention to six sigma? *Quality Progress*, September, 100–103. Milwaukee, WI: American Society for Quality.

Snee, R.D. 2000: Six sigma improves both statistical training and process. *Quality Progress*, October, 68–72. Milwaukee, WI: American Society for Quality.

(g) Statistics – general

Antony, J., Frangou, A. and Kaye, M. 1997: Use your reason to understand statistics. *Quality World*, October, 850–853. London: Institute of Quality Assurance.

ASQ Statistics Division, 1996. *Glossary and Tables for Statistical Quality Control.* Milwaukee, WI: American Society for Quality.

ASQ Statistics Division. 1999: *Improving Performance Through Statistical Thinking*. Milwaukee, WI: American Society for Quality.

Box, G.E.P. 1989: Quality Improvement: An Expanding Domain for the Application of Scientific Method. *Phil. Trans. Roy. Soc.* **A 327**, 617–630

BS 5532 1978: *Statistics–Vocabulary and Symbols*. London: British Standards Institute.

Davis, H.T. 1941: *The analysis of Economic Time Series*. Cowles Commission No. 6. Bloomington, Indiana: Principia Press.

Davis, T.P. 1992: Statistical Methods in Quality Engineering. In: *Proceedings of the Annual Meeting of the British Association for the Advancement of Science*, Southampton, August 23–28.

Morrison, S.J. 1987: SQC is not enough. *The Statistician*, **39**, 5, 439–464. Bury St Edmunds: Institute of Statisticians.

Morrison, S.J. 1992: How (not) to teach statistics. *Quality Forum* **18**, **1**, 5–16. London: Institute of Quality Assurance.

Morrison, S.J. 1998: Looking Back. *Quality World* January, 22–27. Institute of Quality Assurance.

(h) Taguchi methods

Antony, J., Mazharsolook, E. and Kaye, M. 1996: An application of Taguchi's robust parameter design methodology for process improvement. *Quality World Technical Supplement*, March, 35–41. London: Institute of Quality Assurance.

Garzon, I.E., Taher, M.A. and Anderson, A. 2000: Evaluating Taguchi tools through case studies. In: Coleman, S.Y., Stewardson, D. and Fairbairn, L. 2000: *Industrial Statistics in Action*, **11**, 81–97. University of Newcastle upon Tyne.

Park, A. 1991: *Taguchi Methods: Introduction to Quality Engineering*. Dearborn, MI: American Supplier Institute.

Sarin, S. 1997: Teaching Taguchi's approach to parameter design. *Quality Progress*, May 102–106. Milwaukee, WI: American Society for Quality.

Vining, G.G. and Myers, R.H. 1990: Combining Taguchi and response surface philosophies – a dual response approach. *Journal of Quality Technology* **22**, 38–45.

Appendix E

Statistical Tables

Conventional statistical tables, some with fifty (or more) entries recorded up to five decimal places on each page, are unnecessarily detailed for statistical engineering purposes. Buried in such a mass of closely printed numerical data the nature of some of the statistical functions does not become apparent without close scrutiny.

The following abbreviated tables are more revealing. They cover the range of probability values likely to be encountered in engineering situations and they illuminate the statistical methods described in this text in a manner that is easily understood. The values are recorded with sufficient precision for practical purposes and they are presented in a manner that should facilitate understanding of the characteristic nature of each statistical function.

The tabulation of percentage points of the normal distribution identifies with engineering concern about the proportion of individuals outside specification tolerance limits. The tabulation of 5% and 1% critical values of t, F and χ^2 over appropriate data set sizes enables the significance of comparisons to be seen at a glance. The inherent weakness of small data sets in both the t-test and the F-test is clearly indicated.

Percentage points of the normal distribution

P (%)	50	20	10	5	2	1	0.5	0.2	0.1	0.05	0.005	0.0005
X	0.00	0.84	1.28	1.64	2.05	2.33	2.58	2.88	3.09	3.29	3.89	4.42

Statistics for Engineers: an Introduction   S.J. Morrison
© 2009 John Wiley & Sons, Ltd

Percentage points of the F-distribution

$v_1 =$	1		2		5		10		∞	
$P = (\%)$	5	1	5	1	5	1	5	1	5	1
$v_2 = 1$	161.4	4052	199.5	5000	230.2	5764	241.9	6056	254.3	6366
2	18.5	98.5	19.0	99.0	19.3	99.3	19.4	99.4	19.5	99.5
5	6.6	16.3	5.8	13.3	5.1	11.0	4.7	10.1	4.4	9.0
10	5.0	10.0	4.1	7.6	3.3	5.6	3.0	4.9	2.5	3.9
20	4.4	8.1	3.5	5.9	2.7	4.1	2.4	3.4	1.8	2.4
50	4.0	7.2	3.2	5.1	2.4	3.4	2.0	2.7	1.5	1.7
∞	3.8	6.6	3.0	4.6	2.2	3.0	1.8	2.3	1.0	1.0

Percentage points of the t-distribution

v	1	2	5	10	20	50
$P = 5\%$	12.7	4.3	2.6	2.2	2.1	2.0
$P = 1\%$	63.7	9.9	4.0	3.2	2.9	2.7

Percentage points of the χ^2-distribution

P (%)	5	1
$v = 1$	3.8	6.6
2	6.0	9.2
5	11.1	15.1
10	18.3	23.2
20	31.4	37.6
50	67.5	76.2
100	124.8	135.8

Index

The index is in two sections: Engineering and Statistics. The Engineering section identifies the role of engineering technology in the service of industrial quality management. The Statistics section identifies points in the text where statistical terminology is used in an explanatory context.

Statistics for Engineers: an Introduction   S.J. Morrison
© 2009 John Wiley & Sons, Ltd